ENDOCRINOLOGY RESEARCH AND CLINICAL DEVELOPMENTS

ADIPONECTIN

PRODUCTION, REGULATION AND ROLES IN DISEASE

Endocrinology Research and Clinical Developments

Additional books in this series can be found on Nova's website under the Series tab.

Additional e-books in this series can be found on Nova's website under the e-books tab.

ENDOCRINOLOGY RESEARCH AND CLINICAL DEVELOPMENTS

ADIPONECTIN

PRODUCTION, REGULATION AND ROLES IN DISEASE

MASAYOSHI YAMAGUCHI
EDITOR

Nova Science Publishers, Inc.
New York

For permission to use material from this book please contact us:
Telephone 631-231-7269; Fax 631-231-8175
Web Site: http://www.novapublishers.com

Library of Congress Cataloging-in-Publication Data

Adiponectin : production, regulation, and roles in disease / editor, Masayoshi Yamaguchi.
p. ; cm.
Includes bibliographical references and index.
ISBN 978-1-61942-165-3 (hardcover)
I. Yamaguchi, Masayoshi, 1947-
[DNLM: 1. Adiponectin--physiology. WK 185]

615.3'6--dc23

2011042514

Published by Nova Science Publishers, Inc. † New York

CONTENTS

PREFACE

Adipose tissue is considered as an inert energy storage depot for excess calories. Evidence during the past two decades have revolutionized the classic view of adipose tissue, which is now recognized as an active endocrine organ to regulate metabolic homeostasis, cardiovascular physiology and immune functions in the body. Adipose tissue produces a large amount of hormones, growth factors, cytokines, extracellular matrix proteins, and lipid chaperones. The bioactive substances secreted by adipocytes are a group of peptides, collectively termed "adipokines," that possess cytokine-like activities. These peptides have been shown to associate with the symptoms of obesity, such as the metabolic syndrome. Obesity has been recognized as a state of chronic inflammation.

Adiponectin, which is an adipocytokine highly and specifically expressed by adipocytes, belongs to the family of adipokines or enriched hormones. Four independent research groups originally identified adiponectin in the 1990 using different experimental approaches, in both mice and humans. Since its discovery, it attracted much attention particularly in the field of molecular pathophysiology of insulin resistance and type 2 diabete mellitus.

Adiponectin plays an important role in the regulation of food intake and energy homeostasis as well as in vascular homeostasis and immunity. Adiponectin circulates in high and low molecular weight multimeric forms in the bloodstream. Epidemiological evidence has indicated that plasma adiponectin levels are relevant to human health conditions. Transgenic and knockout animal models of adiponectin and its receptors have also confirmed the importance of adiponectin in the metabolic syndrome.

Metabolic syndrome is a combination of metabolic disorders, which includes obesity, insulin resistance, type II diabetes, and cardiovascular disease. The metabolic syndrome has become one of the main causes of mortality in the world. Adiponectin has recently emerged as an anti-inflammatory, anti-atherogenic, and insulin-sensitizing factor providing possible link between white adipose tissue and cardiovascular disease.

Thus, adiponectin plays an important pathophisiological role in the metabolic syndrome. The recent topics concerning production, regulation and roles in disease of adiponectin are focused in this book, and it is composed from chapter 1 to 9.

The role of adiponectin in central nervous system is described in **Chapter I**. Adiponectin has potential anti-diabetic, anti-inflammatory, and anti-atherosclerotic properties. A recent study shows that adiponectin secreted into the circulation can interact with the blood-brain barrier (BBB) and exerts potent central nervous system effects. The influence of adiponectin on BBB function has been demonstrated by measuring its activity on adiponectin receptors.

The expression of adiponectin receptor mRNA has been demonstrated in brain endothelial cells. Adiponectin suppresses pro-inflammatory interleukin-6 release from brain endothelial cells, indicating a modulatory effect of adiponectin on BBB integrity. Using a kainic acid (KA)-induced status epilepticus animal model, it is demonstrated that KA-induced neuronal cell death, BBB leakage, and inflammation are attenuated by adiponectin pretreatment. Adiponectin seems to contribute particularly to the endothelial cells and astrocytes that comprise the BBB, through inhibiting caspase-mediated endothelial cell apoptosis.

The involvement of adiponectin in macrophages and metabolic syndrome is reviewed in **Chapter II**. Metabolic syndrome is a combination of metabolic disorders, which includes obesity, insulin resistance, type II diabetes, and cardiovascular disease. The metabolic syndrome has become one of the main causes of mortality. Macrophages play a role in both metabolic and vascular components of cardiometabolic disease. Recent research findings suggest that alterations in adiponectin effects, solely directed at the macrophage, can explain the co-occurrence of multiple components in the metabolic syndrome trait cluster. The use of macrophages as carriers for adiponectin, a molecule with anti-diabetic and anti-inflammatory functions, provides a novel and innovative approach for prevention and therapy of the metabolic syndrome.

Restenosis after percutaneous coronary intervention (PCI) represents a major limitation to this procedure and despite large efforts, and it is still difficult to stratify patients after PCI according to restenosis risk. A wide range of proteins has been investigated in order to establish new diagnostic/prognostic markers of restenosis after PCI, without major effects. Considering its cardioprotective effects, the role of adiponectin in prediction of adverse cardiovascular events has been studied extensively. However, little attention is paid to adiponectin role in restenosis development after PCI, even though pathophysiology of restenosis covers directly the mechanisms regulated or substantially influenced by adiponectin. **Chapter III** summarizes genetic and pathophysiological relevancies of adiponectin and discusses its value as a biomarker in restenosis diagnosis/prediction.

In excess adipose tissue, the increase in total amount of proinflammatory adipokines, such as tumor necrosis factor-α and interleukin-6 are observed, indicating that obesity is associated with a state of low-grade, chronic inflammation. Adiponectin, however, shows negative correlation with body mass index and decreases its plasma level in obese. As clinical feature, hypoadiponectinaemia is closely related with onset of metabolic syndrome, cardiovascular disease, and type II diabetes. Administration of adiponectin improves insulin resistance and/or inflammatory condition *in vivo*. Thus, adiponectin possessing anti-inflammatory and anti-metabolic dysfunctional characterization is expected to use for treatment and/or preventative agent. In **Chapter IV**, the role of adiponectin in metabolic and immune function, in function of macrophage, is reviewed.

Abnormal circulating adiponectin levels are closely associated with the development of chronic liver diseases, including non-alcoholic hepatic steatosis, steatohepatitis, fibrosis and cirrhosis, as well as hepatocellular carcinoma. The hepatoprotective functions of adiponectin have been demonstrated by numerous clinical, pharmacological, and genetic investigations. The close relationship between fat and liver tissues has been signified by the high prevalence of chronic liver disease associated with the growing pandemic of obesity during past two decade. Under obese conditions, aberrant productions of free fatty acids, pro-inflammatory adipokines and chemokines from adipose tissue are particularly deleterious to the liver tissue, leading to the "adipocentric" concept of hepatic disorder. Numerous adipokines have been found to mediate the adipose-liver cross-talks. **Chapter V** provides an update on how adiponectin, an anti-inflammatory adipokine, acts to protect liver functions, and the therapeutic potential of this protein in chronic liver diseases.

Adiponectin has the potential to correct many metabolic derangements present in sepsis. The altered secretion of adipokines in sepsis is probably connected to the dysfunction adipose tissue undergoes in this disease characterized by reduced insulin sensitivity, inflammation, and metabolic dysfunction reciprocally impairing them. Hypoadiponectinemia in sepsis may be set in a picture of adipose tissue dysfunction, called adiposopathy, and characterized not only by accumulation of fat but also by the inability of adipose tissue to store correctly lipid giving rise to the lipotoxicity and organ damage. Adiponectin may potentially correct many of the metabolic alterations present in sepsis impacting strongly on outcome. This complex of protective activities may explain the relationships between serum adiponectin levels and the outcome in the septic and critically ill patient. The relationships between adipose tissue dysfunction and the related alteration of APN and adipokines secretion, the regulation of inflammation and metabolic dysfunction in sepsis and their prognostic influence are stressed in **Chapter VI**.

Regucalcin is known as a regulatory protein in various intracellular signaling systems. Regucalcin transgenic (TG) rat has been found to induce hyperlipidemia with increasing age; serum triglyceride, high-density lipoprotein-cholesterol, free fatty acid, and albumin are markedly increased in the TG male and female rats with increasing age. The decrease in lipid and glycogen contents in the liver tissues is induced in regucalcin TG rats. The expression of leptin mRNA in the adipose and liver tissues was found to decrease in regucalcin TG rats. Adiponectin mRNA levels were not changed in the adipose tissues of regucalcin TG rats, while its expression was decreased in the liver tissues. These decreases may be involved in the hyperlipidemia induced in regucalcin TG rats. It is proposed that regucalcin is related to hyperlipidemia and that the regucalcin TG rat model may be useful to determine the pathophysiological state and the development of therapeutic tool for hyperlipidemia. **Chapter VII** has been written to outline the recent advances that have been made concerning the role of regucalcin in the regulation of hyperlipidemia induced in regucalcin TG rats with an involvement of adiponectine.

Angiotensin II receptor blockade has been shown to have a beneficial effect on the angiopathies of hypertension and hyperglycemia in patients with type 2 diabete. On the other hand, blood pressure is difficult to control with monotherapy, especially in patients with diabetes, highlighting the requirement of drug contribution such as losartan/ hydrochloro thiazide (LST/HCTZ). However, the effect of LST/HCTZ on the angiopathy-related markers in hypertension and type 2 diabete is poorly understood. The effects of LST/HCTZ on these markers in 82 hypertensive patients with and without type 2 diabete mellitus are introduced in **Chapter VIII**. Findings suggest that LST/HCTZ may be beneficial as an adiponectin-dependent anti-atherosclerotic therapy in patients with type 2 diabete in addition to its anti-hypertensive action.

In pre-eclampsia (PE) after its onset, circulating adiponectin has been reported to increase in the majority but not in all studies. A retrospective analysis was performed with 50 patients who subsequently developed mild, late-onset pre-eclampsia, which is the most frequent manifestation of the disease. These findings are reported in **Chapter IX**. Concentrations of total adiponectin did not differ in sample of first trimester sera between women developing mild, late-onset pre-eclampsia, and matched controls. Adiponectin levels were negatively correlated with the maternal pre-pregnancy weight and with the serum concentrations of leptin and C-reactive protein. Adiponectin does not seem to be a marker for early detection of pre-eclampsia, or at least not for its mild, more frequent form. However, it is interesting to note that it is predominantly the proteins originating from the placenta that show positive correlations with adiponectin.

Recent studies have revealed that adiponectin facilitates insulin action and presents antiatherogenic properties. However, data on the regulation of adipo-nectin plasma concentration and its relationship with abdominal obesity, with hypertriglyceridemia and other metabolic characteristics are still insufficient. Studies, which are reported in **Chapter X,** are to determine the main factors conditioning plasma adiponectin concentration and to evaluate its association with abdominal obesity and biomarkers of insulin resistance and vascular inflammation in a cohort of male adults attending a health care centre. Male adults, at time of a health control, were enrolled in this study. Patients with higher grade of abdominal obesity, in addition to increment in markers of insulin resistance, are found to present higher levels of biomarkers of vascular inflammation and atherosclerosis.

As introduced above, adiponectin plays multifunctional-pathophysiological role in various diseases; neuroprotective effect, involvement in macrophages and metabolic syndrome, role in restenosis development after percutaneous coronary intervention, the role in metabolic and immune function, protective effect on liver functions in chronic liver diseases, an involvement in hyperlipidemia induced in regucalcin transgenic animals, spesis, beneficial effect of losartan/hydrochlorothiazide for an adiponectin-dependent anti-atherosclerotic therapy in patients with type 2 diabete, and role as a biomarker for vascular inflammation and atherosclerosis in early detection of pre-eclampsia. Thus, the novel role of adiponectin is found in recent studies.

The editor believes that this book will help in the research development and understanding of the wide aspects of a pathophysiological role of adiponectin.

Editor Masayoshi Yamaguchi, Ph.D.
Division of Endocrinology and Metabolism and Lipids
Department of Medicine
Emory University School of Medicine
Atlanta, US

In: Adiponectin: Production, Regulation and Roles in Disease ISBN: 978-1-61942-165-3
Editor: Masayoshi Yamaguchi

Chapter I

ROLE OF ADIPONECTIN IN MOUSE BRAIN AFTER SEIZURE

Gu Seob Roh *
Department of Anatomy and Neurobiology,
Institute of Health Sciences, Medical Research Center
for Neural Dysfunction, School of Medicine,
Gyeongsang National University, Republic of Korea

ABSTRACT

Neuronal cell death after excitotoxicity is correlated with blood-brain barrier (BBB) leakage. After seizures, BBB breakdown and inflammation are observed in the hippocampus. This change in BBB integrity is caused by increased local cerebral pressure that occurs during seizures and inflammatory responses such as cytokine production. Seizures are accompanied by increases in angiogenic factors that correlate to the degree of propagation of epileptic activity in animal models. Thus, a better understanding of the angiogenic mechanisms of seizure is critically important.

In general, adiponectin secreted from adipose tissue regulates energy homeostasis and glucose and lipid metabolism in peripheral organs and enhances insulin sensitivity. Adiponectin has potential anti-diabetic, anti-inflammatory, and anti-atherosclerotic properties. Adiponectin is an abundant circulating plasma protein secreted by adipose tissue and exhibits beneficial effects in the vasculature. Circulating adiponectin levels are decreased in patients with obesity, diabetes, and coronary heart disease. Low adiponectin levels are associated with endothelial dysfunction. However, the effects of adiponectin in the brain are still poorly understood.

A recent study showed that adiponectin secreted into the circulation can interact with the BBB and exerts potent central nervous system effects. The influence of adiponectin on BBB function has been demonstrated by measuring its activity on adiponectin receptors. Expression of adiponectin receptor mRNA has been demonstrated in brain endothelial cells. Adiponectin suppresses pro-inflammatory interleukin-6 release from brain endothelial cells, indicating a modulatory effect of adiponectin on BBB integrity. Recent studies have reported that angiogenesis is associated with BBB permeability in epilepsy. The long-lasting increased permeability of the BBB found to be present in

---

* E-mail: anaroh@gnu.ac.kr

various limbic regions, which correlated with seizure activity, suggests a role in increased excitability in the epileptogenic foci that can lead to progression of epilepsy.

This chapter presents the current understanding of the actions of adiponectin in the brain, with particular emphasis on BBB integrity following kainic acid-induced seizure.

Keywords: adiponectin, brain

INTRODUCTION

Adiponectin is an adipose tissue-specific plasma protein that belongs to the family of adipokines or enriched hormones, which play important roles in the regulation of food intake and energy homeostasis as well as in vascular homeostasis and immunity [1,2]. It has been reported that hypoadiponectinemia can be accompanied by an increased severity of obesity and diabetes [3,4]. This negative correlation between obesity and circulating adiponectin has been well established, and adiponectin concentrations increase concomitantly with weight loss [5]. Moreover, plasma adiponectin levels are decreased in patients with cardiovascular diseases, hypertension, or metabolic syndrome [6–8].

In addition to its insulin sensitivity, adiponectin reportedly has anti-inflammatory [9], anti-atherosclerotic [10], and anti-apoptotic [11] effects at the peripheral level. In fact, adiponectin previously was thought to focus in central nervous system (CNS) feeding mechanisms by inhibiting hormone release in the hypothalamus [12]. Several years ago, it was observed that adiponectin contributes to neuroprotection. Jung et al. demonstrated that adiponectin protects SH-SY5Y cells against acetaldehyde-induced cytotoxicity through anti-oxidative and anti-apoptotic pathways [13]. Using a kainic acid (KA)-induced status epilepticus animal model, we demonstrated that KA-induced neuronal cell death, the blood brain barrier (BBB) leakage, and inflammation are attenuated by adiponectin pretreatment [14]. Although this evidence was not directly associated with a correlation between adiponectin and neurons, adiponectin seems to contribute particularly to the endothelial cells and astrocytes that comprise the BBB, through inhibiting caspase-mediated endothelial cell apoptosis. Therefore, this chapter addresses the effects of adiponectin on BBB integrity following KA-induced seizure.

IDENTIFICATION OF ADIPONECTIN

Adiponectin was originally identified by four independent groups using different approaches. Mouse cDNAs for adiponectin, termed Acrp30 [15] and AdipoQ [16], were cloned in mouse 3T3-L1 and 3T3-F442A cells. Human adiponectin is encoded by adipose most abundant gene transcript 1 (apM1) [17] and was purified from plasma as gelatin-binding protein 28 (GBP28) [18]. Circulating adiponectin concentrations are high (500–30,000 ug/L), accounting for approximately 0.01% of total plasma protein [19]. Adiponectin is composed of an N-terminal collagen-like sequence and a C-terminal globular region with significant sequential and structural similarities with the complement factor C1q. A small amount of a processed globular form reportedly was present in human plasma [20]. Adiponectin is induced during adipocyte differentiation, and its secretion is stimulated by insulin. Plasma levels of adiponectin are reduced in obese and diabetic rodents and humans [16,21].

Expression of Adiponectin Receptors

Two adiponectin receptors, adiponectin receptor 1 (AdipoR1) and AdipoR2, have been cloned [3]. Both receptors, which share 67% amino acid identity, appear to be integral membrane receptors containing seven membrane domains. However, contrary to G protein-coupled receptors, their N terminus is intracellular and their C terminus is extracellular [21]. These receptors are expressed widely, including in the brain [22]. AdipoR1 is abundant in muscle and binds with high affinity to globular adiponectin and with low affinity to full-length adiponectin. AdipoR2 is expressed mainly in the liver and has intermediate affinity to both globular and full-length adiponectin. The expression of these two receptors also has been demonstrated in pancreatic β-cells [23], endothelial cells [24], bone-forming cells [25], the hypothalamus [12], adipocytes [26], and the placenta [27]. Moreover, both AdipoR1 and AdipoR2 are present in endothelial cells of the brain [3]. Because these two receptors are expressed in the CNS and the pharmacological responses appear to involve sympathetic activation, adiponectin is considered to affect the central CNS. For example, chronic adiponectin treatment inhibits food intake in obese rats concomitant with weight loss and lowering of glucose and lipids [28].

Effects of Adiponectin in the CNS

Until now, it has been proposed that CNS-mediated effects of adiponectin contribute to weight loss through increased energy expenditure [12]. In mice, intravenous (iv) injection of adiponectin led to an increase in cerebrospinal fluid (CSF) adiponectin, indicating that adiponectin crosses the BBB and the blood-CSF barrier [12]. Two AdipoRs have detected in the brain, especially in the hypothalamus and paraventricular nucleus, supporting this observation [29]. Adiponectin regulates hormone secretion and gene expression in two endocrine cell types of the rat pituitary, somatotrophs and gonadotrophs [30]. In addition to BBB transport, endogenous adiponectin produced within the CNS might have been detected or a fragment of the injected adiponectin could have been immunoreactive. Sensitive radiotracer methods failed to detect significant adiponectin permeation for quantification of BBB influx after iv injection [31,32]. In contrast to rodents, adiponectin was not detected in CSF in healthy humans [32]. Radiolabeled adiponectin does not cross the BBB, but may modify the secretory patterns of BBB cells, whereas no adiponectin is detectable in human CSF. These findings mean it is unlikely that circulating adiponectin has a direct effect on the CNS. However, recent studies using a sandwich enzyme-linked immunosorbent assay (ELISA) system have emphasized that the levels of adiponectin in plasma and CSF are positively correlated [22,33]. Adiponectin mRNA was not found in the human total brain by reverse transcriptase polymerase chain reaction analysis. These data indicate that adiponectin crosses the BBB and/or blood-CSF barrier. Adiponectin was analyzed in plasma and CSF from cognitive normal controls, subjects with mild cognitive impairment, and patients with

Alzheimer's disease (AD), and whether/how adiponectin could relate to the pathogenesis of AD was discussed. The tendency for a higher level of adiponectin in plasma and CSF in AD suggests that this molecule plays a critical role in the onset of AD. Furthermore, it has been speculated that adiponectin enters the brain through the circumventricular organs [34]. In rats, neurons in the area postrema express receptors for adiponectin and show alterations in membrane potentials. For example, microinjection of adiponectin into these circumventricular organs increases arterial blood pressure, suggesting a role for adiponectin in cardiovascular homeostasis by actions in the brain [35,36]. Taken together, these recent studies report biological findings concerning the effects of adiponectin in the CNS.

Effect of Adiponectin on BBB Intergrity in the Brain

Although adipokines such as adiponectin secreted into the circulation can interact with the BBB and exert potent CNS effects [36], adiponectin does not seem to cross the BBB [31]. However, another study demonstrated that systemic injection of adiponectin results in significantly increased levels of adiponectin in serum and CSF [12]. In addition, the influence of adiponectin on BBB function has been demonstrated. Treatment of brain endothelial cells with recombinant adiponectin reduced interleukin (IL)-6 release, indicating that adiponectin has a modulatory effect on BBB integrity [12,32]. Adiponectin has been reported to stimulate endothelial cell growth and angiogenesis [7].

In contrast, another study demonstrated that adiponectin possesses anti-angiogenesis and anti-tumor activities involving caspase-mediated endothelial cell apoptosis [37].

In endothelial cells, adiponectin suppresses the adverse effects of inflammatory cytokines and reduces the oxidative stress induced by oxidized low-density lipoprotein (LDL) [38]. Another study highlighted the anti-inflammatory role of adiponectin signaling by showing that adiponectin blocks the activation of the inhibitor of kappa B-nuclear factor-kappa B (IκB-NF-κB) cascade [39].

BBB Dysfunction in Seizure

The CNS requires a very stable environment to function normally. This stability is provided by the BBB, which isolates the nervous system from the blood. The BBB is an important semi-permeable barrier that protects the brain and spinal cord from harmful substances while permitting gases and nutrients to enter the nervous tissue [40]. The BBB consists of the following structures: endothelial cells, a continuous basement membrane surrounding the capillary, and foot processes of astrocytes [41]. Neuronal damage may be caused by various insults. In particular, ischemic stroke or epilepsy-induced neuronal death is associated with BBB impairment [42,43]. Although there is a good correlation between neuronal death and BBB breakdown, it has not been determined whether either process is dependent on the other. That is, does neuronal death require BBB breakdown, or conversely, does neuronal death significantly contribute to leakage of plasma proteins?

This chapter addresses the dependence of seizure-induced neuronal death on BBB leakage. Recent studies reported that BBB leakage occurs during epileptogenesis and the chronic epileptic phase, and can then contribute to the progress of epilepsy [44]. Changes in BBB integrity may be caused by increased local blood pressure during seizures [45], free radical formation [46], inflammatory responses such as leukocyte recruitment and cytokine and interleukin production [47], and/or the loss of tight junction molecules [48].

Brain endothelial cells are capable of expressing several adhesion molecules, including intercellular adhesion molecule-1 (ICAM-1), ICAM-2, and vascular cell adhesion molecule-1 (VCAM-1) [49]. The promotion of leukocyte adhesion to the cerebral endothelium may be mediated through a modulation of these adhesion molecules, which then permits the migration of lymphocytes across the BBB via a transcellular route [50].

In general, the rapid secretion of growth factors after seizure is considered to be a protective mechanism. Recent studies have reported that angiogenesis is associated with BBB permeability in temporal lobe epilepsy [44,51].

This is the case for vascular endothelial growth factor (VEGF) release; its anti-epileptogenic and neuroprotective properties were recently demonstrated both in vivo and in vitro [52,53]. VEGF is a key regulator of angiogenesis, vascular permeability of the endothelium, and the survival of endothelial cells [54]. Increases in VEGF contribute to the BBB breakdown and inflammation observed after seizure [55]. In rats, increases in VEGF in neurons and glia in the hippocampus and limbic cortex were described after status epilepticus [56]. VEGF was observed in astrocytes in temporal lobe epilepsy with hippocampal sclerosis [51].

The extravasation of immunoglobin (Ig) Gs and other serum proteins that accumulate in the cytoplasm of neurons has been reported in the case of acute BBB breakdown and auto-immune and neurodegenerative diseases [57,58]. We confirmed BBB impairment in KA-treated mice using IgG immunohistochemistry, which has been used to monitor protein extravasation and disruption of the BBB [14,44].

One consequence of BBB leakage is the accumulation of serum proteins that enter the brain, which may contribute to increased excitability [59]. Furthermore, there is other evidence that astrocytes, which are the major component of the BBB, are adversely affected by excitotoxins [60]. After KA injection, a loss of glial fibrillary acidic protein (GFAP) coincides with IgG extravasation [61]. Even if angiogenesis accounts for neuroprotective mechanisms during seizures, angiogenesis may be involved in chronic permeability of the BBB, with severe consequences such as neurovascular uncoupling, inflammation, and excitability [51].

Nitric oxide (NO) plays an important role in the regulation of cerebral circulation, both in basal conditions and during various types of neuronal activation [62]. NO synthesis is well known to have a key regulatory function in endothelial cells. NO participates in the cerebrovascular response to both generalized seizures induced by systemic KA [63] and focal seizures induced by bicuculline [64].

In eNOS-/-mice, no local cerebral blood flow (LCBF) increases were recorded within the epileptic focus, and generalized contralateral LCBF decreases were noticed in two-thirds of the brain areas compared with saline-injected eNOS-/-mice [64]. This study addressed the apparent involvement of eNOS in the regulation of the vascular response both within and at a distance from the epileptic focus.

Effective Mechanism of Adiponectin in Mouse Hippocampus After Seizure

Some studies have revealed that adiponectin has potential anti-inflammatory, anti-diabetic, and anti-apoptotic properties [8]. Adiponectin has been reported to inhibit acetaldehyde-induced apoptosis of the human neuroblastoma cell line SH-SY5Y [13]. Acetaldehyde, a metabolite of alcohol, has been shown to selectively damage mitochondria when combined with 1-methyl-4-phenyl-2,3,6-tetrahydropyridine [65]. Although AdipoR2 is expressed in SH-SY5Y cells, its expression level does not change with acetaldehyde and adiponectin treatment. These results show that adiponectin decreases acetaldehyde-induced cell death and reactive oxygen species production in SH-SY5Y cells by inducing anti-oxidant enzymes. These data are consistent with those of a previous report showing that adiponectin suppresses superoxide generation and enhances eNOS activity in endothelial cells treated with oxidized LDL [24]. Adiponectin also inhibited acetaldehyde-induced caspase-3 cleavage-like cytochrome c release. Thus, the data suggested that adiponectin inhibits loss of cell viability by overcoming oxidative stress and regulating apoptotic genes. Based on these findings, we hypothesize that adiponectin could protect neurons against excitotoxicity-induced cell death. Therefore, we have demonstrated for the first time that KA-induced hippocampal cell death was attenuated by intracerebroventricular (icv) injection of adiponectin [14]. The data showed that KA-induced BBB leakage and VEGF, eNOS, and NF-κB expression in the hippocampus were attenuated by icv injection of recombinant adiponectin.

It is generally known that KA-induced neuronal damage is involved in oxidative stress-mediated apoptosis, including BBB leakage, inflammation, and excitotoxicity. Although a good correlation exists between BBB leakage and inflammation in the hippocampus following KA-induced seizure, it has not been determined whether either process is dependent on the other. Thus, we investigated the effect of adiponectin pretreatment on BBB leakage and neuroprotection in the hippocampus after seizure. In KA-induced status epilepticus, there was no difference in seizure activity between KA-treated mice with or without adiponectin. CSF adiponectin in humans and rodents was approximately 0.1% of the serum concentration, and adiponectin mRNA was not expressed in the brain [22]. Circulating adiponectin levels are decreased in metabolic syndrome [66]. Low adiponectin levels are also associated with vascular injury and atherosclerosis [67]. A recent study reported that hypoadiponectinemia is linked to valvular inflammation and more rapid progression of heart disease [68]. Based on ELISA results, serum adiponectin was markedly decreased in the KA-treated group, whereas the KA-treated mice with adiponectin had serum adiponectin concentrations comparable to those of the control group [14]. However, the expression of AdipoR1 was increased in KA-treated mice compared with control mice. There was no difference in AdipoR2 protein expression between control and KA-treated mice. These data indicate that the increased hippocampal AdipoR1 expression in KA-treated mice may be a response to low serum adiponectin. Thus, low adiponectin may aggravate KA-induced BBB dysfunction in the hippocampus [14].

Recent studies have reported that BBB leakage in epilepsy is associated with angiogenesis [44,51]. The increased permeability of the BBB suggests a role in increased excitability in epileptogenic foci [44]. Two days after systemic KA injection we confirmed

BBB leakage in the KA-treated mice using immunohistochemistry for IgG, which has been used to monitor protein extravasation and disruption of the BBB [14,44]. One consequence of BBB leakage is the accumulation of serum proteins that enter the brain, which may contribute to increased excitability [59]. Therefore, the data indicate that adiponectin may contribute to inhibition of serum IgG proteins from KA-induced BBB breakdown.

Dysfunction of angiogenesis is a characteristic of metabolic syndrome, which includes atherosclerosis, obesity, diabetes, and hypertension [7,69]. However, an interaction between adiponectin and angiogenesis has not been elucidated in the CNS. Another study demonstrated that adiponectin possesses anti-angiogenesis activities involving caspase-mediated endothelial cell apoptosis [37]. VEGF has dramatic effects on brain vasculature and can promote the BBB breakdown and inflammation observed after seizure [56]. Oxidative stress-induced VEGF in endothelial cells was suppressed by adiponectin [70]. In our previous study, we found that adiponectin attenuated KA-induced VEGF expression in the hippocampus [14].

eNOS is the mediator of the cerebrovascular response within the epileptic focus [64]. The antioxidant curcumin attenuated eNOS expression in the hippocampus after KA treatment [71]. Immunohistochemistry showed that adiponectin attenuated the KA-induced eNOS expression in the hippocampus. This result suggests that the reduction in eNOS may enhance the protective effects of adiponectin on BBB function. Interestingly, adiponectin has been shown to exert putative anti-atherogenic properties by suppressing tumor necrosis factor (TNF)-α induced NF-κB activation through a cyclic adenosine monophosphate (cAMP)-dependent pathway [7]. In endothelial cells, adiponectin suppresses inflammatory cytokines and reduces the oxidative stress induced by oxidized low density lipoprotein (LDL) [24]. Adiponectin blocks activation of the IkB-NF-κB cascade [39]. There is evidence to suggest that NF-κB plays a dynamic role in the degenerative process after brain injuries. In particular, KA treatment induced activation of NF-κB in degenerating neurons over the following 24 h [72]. Immunohistochemistry showed that adiponectin attenuated the KA-induced increase in NF-κB-positive cells in hippocampal CA3 regions. Therefore, the data raise the intriguing possibility that adiponectin-mediated control of NF-κB activation may modulate the process of inflammation in the hippocampus after KA treatment.

CONCLUSION

This chapter highlights the functional role of adiponectin in the mouse hippocampus following KA-induced seizure. Until now, an understanding of the mechanisms of action of various adipokines or hormones secreted by adipose tissue in peripheral levels was considered necessary to understand the functional role of adiponectin in human diseases. The data of a recent study support the hypothesis that icv injection of recombinant adiponectin protects neurons from BBB breakdown and inflammation. In addition, the ability to manipulate serum and CSF adiponectin levels with recombinant adiponectin may lead to new insights into neuronal survival and microvessels within the brain. Therefore, this chapter presents the possibility that adiponectin in the CNS may attenuate excitotoxicity-induced oxidative stress, inflammation, and angiogenesis in the BBB and subsequently protect KA-induced hippocampal cell death.

REFERENCES

[1] M.W. Rajala, S. Obici, P.E. Scherer, and L. Rossetti L, *J. Clin. Invest.* 111, 225 (2003).
[2] P. Trayhurn and I.S. Wood, *Biochem. Soc. Trans.* 33, 1078 (2005).
[3] T. Yamauchi, J. Kamon, Y. Ito, A. Tsuchida, T. Yokomizo, S. Kita, T. Sugiyama, M. Miyagishi, K. Hara, M. Tsunoda, K. Murakami, T. Ohteki, S. Uchida, S. Takekawa, H. Waki, N.H. Tsuno, Y. Shibata, Y. Terauchi, P. Froguel, K. Tobe, S. Koyasu, K. Taira, T. Kitamura, T. Shimizu, R. Nagai, and T. Kadowaki, *Nature* 423, 762 (2003).
[4] Y. Arita, S. Kihara, N. Ouchi, M. Takahashi, K. Maeda, J. Miyagawa, K. Hotta, I. Shimomura, T. Nakamura, K. Miyaoka, H. Kuriyama, M. Nishida, S. Yamashita, K. Okubo, K. Matsubara, M. Muraguchi, Y. Ohmoto, T. Funahashi, and Y. Matsuzawa, *Biochem. Biophys. Res. Commun.* 257, 79 (1999).
[5] M. Faraj, P.J. Havel, S. Phélis, D. Blank, A.D. Sniderman, and K. Cianflone, *J. Clin. Endocrinol. Metab.* 88, 1594 (2003).
[6] M. Kumada, S. Kihara, S. Sumitsuji, T. Kawamoto, S. Matsumoto, N. Ouchi, Y. Arita, Y. Okamoto, I. Shimomura, H. Hiraoka, T. Nakamura, T. Funahashi, Y. Matsuzawa, and Osaka CAD Study Group, *Arterioscler. Thromb. Vasc. Biol.* 23, 85 (2003).
[7] N. Ouchi, H. Kobayashi, S. Kihara, M. Kumada, K. Sato, T. Inoue, T. Funahashi, and K. Walsh, *J. Biol. Chem.* 279, 1304 (2004).
[8] M.E. Trujillo and P.E. Scherer, *J. Intern. Med.* 257, 167 (2005).
[9] Y. Hattori, Y. Nakano, S. Hattori, A. Tomizawa, K. Inukai, and K. Kasai, *FEBS Lett.* 582, 1719 (2008).
[10] X.J. Cai, C.J. Li, L. Chen, Y.Y. Rong, Y. Zhang, and M. Zhang, *Med. Hypotheses* 70, 1044 (2008).
[11] M. Park, B. Youn, X.L. Zheng, D. Wu, A. Xu, and G. Sweeney, *PLoS One* 6, e19143 (2011).
[12] Y. Qi, N.N. Takahashi, S.M. Hileman, H.R. Patel, A.H. Berg, U.B. Pajvani, P.E. Scherer, and R.S. Ahima, *Nat. Med.* 10, 524 (2004).
[13] T.W. Jung, J.Y. Lee, W.S. Shim, E.S. Kang, J.S. Kim, C.W. Ahn, H.C. Lee, and B.S. Cha, *Biochem. Pharmacol.* 72, 616 (2006).
[14] B.T. Jeon, H.J. Shin, J.B. Kim, Y.K Kim, D.H. Lee, K.H. Kim, H.J. Kim, S.S. Kang, G.J. Cho, W.S. Choi, and G.S. Roh, *Brain Res. Rev.* 61, 81 (2009).
[15] P.E. Scherer, S. Williams, M. Fogliano, G. Baldini, and H.F. Lodish, *J. Biol. Chem.* 270, 26746 (1995).
[16] E. Hu, P. Liang, and B.M. Spiegelman, *J. Biol. Chem.* 271, 10697 (1996).
[17] K. Maeda, K. Okubo, I. Shimomura, T. Funahashi, Y. Matsuzawa, and K. Matsubara, *Biochem. Biophys. Res. Commun.* 221, 286 (1996).
[18] Y. Nakano, T. Tobe, N.H. Choi-Miura, T. Mazda, and M. Tomita, *J. Biochem.* 120, 803 (1996).
[19] A.H. Berg, T.P. Combs, and P.E. Scherer, *Trends Endocrinol. Metab.* 13, 84 (2002).
[20] J. Fruebis, T.S. Tsao, S. Javorschi, D. Ebbets-Reed, M.R. Erickson, F.T. Yen, B.E. Bihain, and H.F. Lodish, *Proc. Natl. Acad. Sci. U. S. A.* 98, 2005 (2001).
[21] T. Kadowaki and T. Yamauchi, *Endocr. Rev.* 26, 439 (2005).

[22] M. Neumeier, J. Weigert, R. Buettner, J. Wanninger, A. Schäffler, A.M. Müller, S. Killian, S. Sauerbruch, F. Schlachetzki, A. Steinbrecher, C. Aslanidis, J. Schölmerich, and C. Buechler, *Am. J. Physiol. Endocrinol. Metab.* 293, E965 (2007).

[23] Kharroubi, J. Rasschaert, D.L. Eizirik, and M. Cnop, *Biochem. Biophys. Res. Commun.* 312, 1118 (2003).

[24] H. Motoshima, X. Wu, K. Mahadev, and B.J. Goldstein, *Biochem. Biophys. Res. Commun.* 315, 264 (2004).

[25] H.S. Berner, S.P. Lyngstadaas, A. Spahr, M. Monjo, L. Thommesen, C.A. Drevon, U. Syversen, and J.E. Reseland, *Bone* 35, 842 (2004).

[26] L. Nilsson, N. Binart, M. Bohlooly-Y, M. Bramnert, E. Egecioglu, J. Kindblom, P.A. Kelly, J.J. Kopchick, C.J. Ormandy, C. Ling, and H. Billig, *Biochem. Biophys. Res. Commun.* 331, 1120 (2005).

[27] J. Chen, B. Tan, E. Karteris, S. Zervou, J. Digby, E.W. Hillhouse, M. Vatish, and H.S. Randeva, *Diabetologia* 49, 1292 (2006).

[28] S. Shklyaev, G. Aslanidi, M. Tennant, V. Prima, E. Kohlbrenner, V. Kroutov, M. Campbell-Thompson, J. Crawford, E.W. Shek, P.J. Scarpace, and S. Zolotukhin, *Proc. Natl. Acad. Sci. U. S. A.* 100, 14217 (2003).

[29] K. Kos, A.L. Harte, N.F. da Silva, A. Tonchev, G. Chaldakov, S. James, D.R. Snead, B. Hoggart, J.P. O'Hare, P.G. McTernan, and S. Kumar, *J. Clin. Endocrinol. Metab.* 92, 1129 (2007).

[30] F. Rodriguez-Pacheco, A.J. Martinez-Fuentes, S. Tovar, L. Pinilla, M. Tena-Sempere, C. Dieguez, J.P. Castaño, and M.M. Malagon, *Endocrinology* 148, 401 (2007).

[31] W. Pan, H. Tu, and A.J. Kastin, *Peptides* 27, 911 (2006).

[32] J. Spranger, S. Verma, I. Gohring, T. Bobbert, J. Seifert, and A. Sindler, *Diabetes* 55, 141 (2006).

[33] K. Une, Y.A. Takei, N. Tomita, T. Asamura, T. Ohrui, K. Furukawa, and H. Arai, *Eur. J. Neurol.* 18, 1006 (2011).

[34] R.S. Ahima, Y. Qi, N.S. Singhal, M.B. Jackson, and P.E. Scherer, *Diabetes* 55, S145 (2006).

[35] M. Fry, P.M. Smith, T.D. Hoyda, M. Duncan, R.S. Ahima, K.A. Sharkey, and A.V. Ferguson, *J. Neurosci.* 26, 9695 (2006).

[36] W. Pan and A.J. Kastin, *Peptides* 28, 1317 (2007).

[37] E. Bråkenhielm, N. Veitonmäki, R. Cao, S. Kihara, Y. Matsuzawa, B. Zhivotovsky, T. Funahashi, and Y. Cao, *Proc. Natl. Acad. Sci. U. S .A.* 101, 2476 (2004).

[38] H. Motoshima, X. Wu, K. Mahadev, and B.J. Goldstein, *Biochem. Biophys. Res. Commun.* 315, 264 (2004).

[39] C. Kobashi, M. Urakaze, M. Kishida, E. Kibayashi, H. Kobayashi, S. Kihara, T. Funahashi, M. Takata, R. Temaru, A. Sato, K. Yamazaki, N. Nakamura, and M. Kobayashi, *Circ. Res.* 97, 1245 (2005).

[40] N.J. Abbott, *Cell Mol. Neurobiol.* 25, 5 (2005).

[41] N.J. Abbott, L. Rönnbäck, and E. Hansson, *Nat. Rev. Neurosci.* 7, 41 (2006).

[42] C. Nitsch and I. Klatzo, *J. Neurol. Sci.* 59, 305 (1983).

[43] Z.G. Zhang, L. Zhang, Q. Jiang, R. Zhang, K. Davies, C. Powers, N. Bruggen, and M. Chopp, *J. Clin. Invest.* 106, 829 (2000).

[44] E.A. van Vliet, S. da Costa Araújo, S. Redeker, R. van Schaik, E. Aronica, and J.A. Gorter, *Brain* 130, 521 (2007).

[45] C. Nitsch, and I. Klatzo, *J. Neurol. Sci.* 59, 305 (1983).
[46] B. Oztaş, H. Koçak, P. Oner, and M. Küçük, *J. Neurosci. Res.* 62, 750 (2000).
[47] Vezzani and T. Granata, *Epilepsia* 46, 1724 (2005).
[48] P. Ballabh, A. Braun, and M. Nedergaard, *Neurobiol. Dis.* 16, 1 (2004).
[49] B. Engelhardt, *J. Neural Transm.* 113, 477 (2006).
[50] E. Oby and D. Janigro, *Epilepsia* 47, 1761 (2006).
[51] V. Rigau, M. Morin, M.C. Rousset, F. de Bock, A. Lebrun, P. Coubes, M.C. Picot, M. Baldy-Moulinier, J. Bockaert, A. Crespel, and M. Lerner-Natoli, *Brain* 130, 1942 (2007).
[52] D.P. McCloskey, S.D. Croll, and H.E. Scharfman, *J. Neurosci.* 25, 8889 (2005).
[53] J.N. Nicoletti, S.K. Shah, D.P. McCloskey, J.H. Goodman, A. Elkady, H. Atassi, D. Hylton, J.S. Rudge, H.E. Scharfman, and S.D. Croll, *Neuroscience* 151, 232 (2008).
[54] N. Ferrara, H.P. Gerber, and J. LeCouter, *Nat. Med.* 9, 669 (2003).
[55] H.E. Scharfman, J.H. Goodman, A.L. Sollas, and S.D Croll, *Exp. Neurol.* 174, 201 (2002).
[56] S.D. Croll, J.H. Goodman, and H.E. Scharfman, *Adv. Exp. Med. Biol.* 548, 57 (2004).
[57] R.H. Fabian and G. Petroff, *Neurology* 37, 1780 (1987).
[58] K. Kitz, H. Lassmann, D. Karcher, and A. Lowenthal, *Acta Neuropathol.* 63, 41 (1984).
[59] E. Seiffert, J.P. Dreier, S. Ivens, I. Bechmann, O. Tomkins, U. Heinemann, and A. Friedman, *J. Neurosci.* 24, 7829 (2004).
[60] R.C. Janzer and M.C. Raff, 1987. *Nature* 325, 253 (1987).
[61] S.J. Bolton and V.H. Perry, 1998. *Exp. Neurol.* 154, 231 (1998).
[62] N. Akgören, M. Fabricius, and M. Lauritzen, *Proc. Natl. Acad. Sci. U. S. A.* 91, 5903 (1994).
[63] C. Montécot, J. Borredon, J. Seylaz, and E. Pinard, *J. Cereb. Blood Flow Metab.* 17, 94 (1997).
[64] Pereira de Vasconcelos, R.A. Baldwin, and C.G. Wasterlain, *Proc. Natl. Acad. Sci. U. S. A.* 92, 3175 (1995).
[65] G.U. Corsini, S. Pintus, A. Bocchetta, M.P. Piccardi, and M. Del Zompo. *J. Pharmacol. Exp. Ther.* 238, 648 (1986).
[66] S. Furukawa, T. Fujita, M. Shimabukuro, M. Iwaki, Y. Yamada, Y. Nakajima, O. Nakayama, M. Makishima, M. Matsuda, and I. Shimomura, *J. Clin. Invest.* 114, 1752 (2004).
[67] N. Kubota, Y. Terauchi, T. Yamauchi, T. Kubota, M. Moroi, J. Matsui, K. Eto, T. Yamashita, J. Kamon, H. Satoh, W. Yano, P. Froguel, R. Nagai, S. Kimura, T. Kadowaki, and T. Noda, *J. Biol. Chem.* 277, 25863 (2002).
[68] D. Mohty, P. Pibarot, N. Côté, A. Cartier, A. Audet, J.P. Després, and P. Mathieu, *Cardiology* 118, 140 (2011).
[69] P. Carmeliet, *Nat. Med.* 9, 653 (2003).
[70] K. Mahadev, X. Wu, S. Donnelly, R. Ouedraogo, A.D. Eckhart, B.J. Goldstein, *Cardiovasc. Res.* 78, 376 (2008).
[71] H.J. Shin, J.Y. Lee, E. Son, D.H. Lee, H.J. Kim, S.S. Kang, G.J. Cho, W.S. Choi, and G.S. Roh, *Neurosci. Lett.* 416, 49 (2007).
[72] S.J. Won, H.W. Ko, E.Y. Kim, E.C. Park, K. Huh, N.P. Jung, I. Choi, Y.K. Oh, H.C. Shin, and B.J. Gwag, *Neuroscience* 94, 83 (1999).

In: Adiponectin: Production, Regulation and Roles in Disease ISBN: 978-1-61942-165-3
Editor: Masayoshi Yamaguchi

Chapter II

ADIPONECTIN, MACROPHAGE, AND METABOLIC SYNDROME

Yuchang Fu*
University of Alabama at Birmingham,
Birmingham, Alabama, US

ABSTRACT

Metabolic syndrome is a combination of metabolic disorders, which includes obesity, insulin resistance, type II diabetes, and cardiovascular disease. The metabolic syndrome has become one of the main causes of mortality in the world. Changing one's lifestyle is currently the preferred treatment for preventing and curing the metabolic syndrome. Therefore, finding an effective way to clinically treat the metabolic syndrome is a big challenge.

Adiponectin is one adipokine which is mainly expressed and secreted from adipose tissue. Adiponectin circulates in high and low molecular weight multimeric forms in the bloodstream. Epidemiological evidence has indicated that plasma adiponectin levels are relevant to human health conditions. Transgenic and knockout animal models of adiponectin and its receptors have also confirmed the importance of adiponectin in the metabolic syndrome.

Macrophages are a heterogeneous and plastic population of phagocytic cells. Macrophage cells can enter other tissues through the bloodstream, and interact with the other cells in their resident tissues. Recent research results suggest that alterations in adiponectin effects in macrophages can improve metabolic syndrome in animal models. The use of macrophages as carriers for enhancing adiponectin functions in other metabolic tissues probably provides a novel and innovative approach for prevention and therapy of the metabolic syndrome.

Keywords: adiponectin, macrophage

---

* E-mail: yfu@uab.edu

Introduction

Metabolic syndrome, defined by multiple metabolic disorders including obesity, insulin resistance, type II diabetes, and cardiovascular disease, has become increasingly common in the United States. It's estimated that over 50 million Americans have these diseases. Since metabolic syndrome is a cluster of conditions that occur together, except of aggressive lifestyle changes and treatment of each component of the metabolic syndrome separately, taking on all of them together might seem overwhelming and impossible; lifestyle modification is currently the preferred treatment for metabolic syndrome. Thus, finding an effective way to clinically treat the multiple risk factors linked to the metabolic syndrome is a big challenge for us.

Adiponectin (also known as apM1, AdipoQ, Gbp28 and Acrp30) is one such adipokine, secreted only from adipocytes, which circulates in high and low molecular weight multimeric forms. Epidemiological evidence has indicated that plasma adiponectin levels are reduced in patients with the metabolic syndrome. In the circulation, adiponectin exerts bioeffects on multiple cell types and has insulin-sensitizing, anti-inflammatory, and anti-atherosclerotic properties. Accordingly, administration of adiponectin to intact rodents improved the metabolic syndrome. Adiponectin and its receptor transgenic and knockout mouse models have confirmed the importance of adiponectin in the metabolic syndrome.

Macrophages are a heterogeneous and plastic population of phagocytic cells, which arise from circulating myeloid-derived blood monocytes, entering target tissues, and gaining phenotypic and functional attributes partly determined by their tissue of residence. Macrophages play a role in both metabolic and vascular components of cardiometabolic disease. Recent research results suggest that alterations in adiponectin effects, solely directed at the macrophage, can explain the co-occurrence of multiple components in the metabolic syndrome trait cluster and provide a novel unifying explanation for the pathophysiology of metabolic syndrome. The use of macrophages as carriers for adiponectin, a molecule with anti-diabetic and anti-inflammatory functions, provides a novel and innovative approach for prevention and therapy of the metabolic syndrome.

Metabolic Syndrome and Its Treatment

The metabolic syndrome (also known as syndrome X or the dysmetabolic syndrome) is a combination of medical disorders that exist in a single individual, including central obesity, dyslipidemia characterized by high blood triglycerides and low high-density lipoprotein (HDL) cholesterol levels, hypertension, and hyperglycemia (high blood glucose levels). Without treatments of these medical disorders, the metabolic syndrome would develop into serious diseases, such as type II diabetes and cardiovascular disease. The prevalence of metabolic syndrome among adults in the United States is estimated to be ~25% and it is considered one of the main causes of mortality in the United States and in developing countries [1–3].

Since the metabolic syndrome is a cluster of conditions that occur together, except of aggressive lifestyle changes and treatment of each component, such as hyperlipidemia, high blood pressure and high blood glucose, of the metabolic syndrome separately, taking on all of them together might seem overwhelming and impossible; lifestyle modification is currently the preferred treatment of metabolic syndrome. Thus, finding an effective way to clinically treat the multiple risk factors linked to the metabolic syndrome is a big challenge for us.

In addition to changes in lifestyle, medications to control each component of the metabolic syndrome, such as cholesterol levels, lipid levels, and high blood pressure have been used as the complementally treating methods with the lifestyle modification. However, since some medications used in the metabolic syndrome often have more than one effect, either beneficial or harmful, to several components in metabolic syndrome, correct selection of the medications for treating metabolic syndrome is an important consideration when discussing the choice medication for treating the metabolic syndrome. For example, a class of drugs called thiazolidinediones (TZDs) including pioglitazone (Actos), rosiglitazone (Avandia) and troglitazone (Rezulin) has been used to lower blood glucose levels to achieve glycemic goals in metabolic syndrome or diabetes. However, due to a reported increase in heart attack (rosiglitazone) and serious liver toxicity (troglitazone) in patients taking these two drugs, the United States FDA (Food and Drug Administration) has removed Avandia and Rezulin from the market based on their harmful side effects on physiological metabolism [4, 5]. Recently, even the sale of Actos (pioglitazone) has been suspended in European markets based on its potential increased risk of bladder cancer in patients using pioglitazone for treatment of diabetes [6]. Although at this time, the FDA has not concluded that Actos increases the risk of bladder cancer, in anticipation, an updated formal recommendation from the FDA about Actos will be coming in the very near future. Thus, the limitation of clinical utilization of TZDs highlights the need for additional alternative medications that are both effective and safe for the prevention and treatment of metabolic syndrome.

ADIPONECTIN AND ITS FUNCTIONS

Adiponectin [7] is a 244 amino acid protein which has four distinct regions, and it shows similarity to the complement factor C1q family of proteins. The first region of adiponectin protein is a short signal sequence that targets the hormone for secretion outside the cell; the second region is a short sequence that varies between species with no homology to other proteins; the third region is a 65 amino acid region involved in collagen-triple helix formation with similarity to collagenous proteins; the last region of adiponectin protein is a globular domain at the carboxyl-terminus which shows a very high degree of homology with subunits of C1q and the globular domains of type VIII and type X collagens.

Adiponectin, also known as apM1 (adipose most abundant gene transcript 1) [8], AdipoQ (adipose gene Q) [9], Gbp28 (gelatin-binding protein of 28 kDa) [10], and Acrp30 (adipocyte complement-related protein of 30 kDa) [11], is an adipokine identified by screening adipose-specific genes in the human cDNA projects [10]. Adiponectin expresses and secretes from adipocytes, and circulates in high (hexamers or larger) and low (trimers or smaller) molecular weight multimeric forms in bloodstream [12-14]. Adiponectin is very abundant in plasma (2-20 mg/mL) relative to many other adipokines. Epidemiological evidence has indicated that circulating adiponectin levels are reduced in patients with insulin resistance, type II diabetes, obesity, or cardiovascular disease [15–17]. When examined, low plasma adiponectin levels in these metabolic disease states are accompanied by reduced adiponectin gene expression in adipose tissue [18, 19]. There is also evidence that adiponectin gene polymorphisms may be associated with hypoadiponectinemia together with insulin resistance and type II diabetes [20]. Also, low levels of adiponectin and adiponectin gene variation have been associated with obesity, type II diabetes, and cardiovascular disease. Conversely, in the circulation, adiponectin exerts bioeffects on multiple cell types and has insulin-sensitizing, anti-inflammatory, and anti-atherosclerotic properties, therefore, increased circulating levels of adiponectin are associated with improvement in the metabolic syndrome.

The metabolic effects and action mechanisms of adiponectin on metabolic syndrome are interesting research topics and recent studies have established our better understanding of the functions of adiponectin in metabolic syndrome and diseases.

As a cell model for adipose tissue, 3T3-L1 cells, which express and secrete abundant adiponectin, have been broadly used in biological research for studying the adipocyte-like phenotypes [21]. When adiponectin gene is overexpressed in the 3T3-L1 adipocytes, it exerts as an autocrine factor in these adipocytes for promoting cell proliferation and differentiation from preadipocytes into adipocytes, augmenting programmed gene expression responsible for adipogenesis, and increasing lipid content and insulin responsiveness of the glucose transport system in adipocytes [22]. Similar studies have been observed in adiponectin adipocyte-specific transgenic mice, in which adiponectin gene are specifically overexpressed in adipose tissues with an aP2 (adipocyte fatty acid binding protein, FABP4) enhancer/promoter sequence [23]. In these adiponectin transgenic mice, there is a significant increase in percent total body fat, which reached approximately 30% of their total body weight, in contrast to 18% in their wild-type siblings. These transgenic mice with two- to three fold elevated adiponectin levels are protected against high fat diet (HFD)-induced insulin resistance. When these adiponectin transgenic mice have been crossed with the diabetic *ob/ob* mice which lack leptin and display hyperglycemia, hyperinsulinemia, and dyslipidemia, overexpressing adiponectin in these mice has completely rescued the diabetic phenotype in *ob/ob* mice with dramatically improved glucose as well as positively affected serum triglyceride levels even these crossed transgenic mice are morbidly obese with significantly higher levels of adipose tissue than their *ob/ob* littermates [24]. Recent studies have further shown overexpressed adiponectin in adipose tissue as a factor to increase the metabolic flexibility of the adipose tissue, enhancing its ability to maintain proper function under metabolically challenging conditions [25].

On the other hand, it has been demonstrated that adiponectin knockout mice develop insulin resistance, glucose intolerance, hyperglycemia and hypertension, all characteristics of metabolic syndrome [26–28]. These adiponectin knockout mice have shown moderate insulin resistance with glucose intolerance despite body weight gain similar to that of wild-type mice and 2-fold more neointimal formation in response to external vascular cuff injury than wild-type mice, suggesting that adiponectin plays a protective role against insulin resistance and atherosclerosis in vivo. When examining these adiponectin knockout mice in details, there is also evidence to show delayed clearance of free fatty acid in plasma, low levels of fatty-acid transport protein 1 (FATP-1) mRNA in muscle, high levels of tumor necrosis factor-alpha (TNF-α) mRNA in adipose tissue and high plasma TNF-□α concentrations in these knockout mice. These knockout mice also exhibit severe diet-induced insulin resistance with reduced insulin-receptor substrate 1 (IRS-1)-associated phosphatidylinositol 3 kinase (PI3K) activity in muscle. More importantly, viral mediated adiponectin expression in these knockout mice can reverse some phenotypes in these knockout mice. These studies have further revealed important roles of adiponectin in regulating inflammatory response and glucose/lipid metabolism related to the metabolic syndrome.

Two cell-surface trans-membrane receptors, named as AdipoR1 and AdipoR2, for adiponectin binding have been identified by expression cloning [29]. These two adiponectin receptors are predicted to contain seven transmembrane domains with an extracellular carboxy terminus and an intracellular amino terminus, a structure that is opposite to that of all other G-coupled protein receptors. Expression of AdipoR1/R2 or suppression of AdipoR1/R2 expression by small-interfering RNA (siRNA) experiments indicates that AdipoR1 is a high-affinity receptor for globular adiponectin and a low-affinity receptor for full-length adiponectin, whereas AdipoR2 has an intermediate affinity for both the full-length and globular species. Both of adiponectin receptors are ubiquitously expressed; but AdipoR1 is abundantly expressed in skeletal muscle, whereas AdipoR2 is predominantly expressed in the liver. And in muscle and liver cells, signal transduction involves the phosphorylation and activation of AMP-activated kinase (AMPK), which plays a pivotal role in nutrient sensing and substrate metabolism, AdipoR1/R2 mediate increased AMP kinase and peroxisome proliferator-activated receptor alpha (PPAR-α) ligand activities, as well as fatty-acid oxidation and glucose uptake by adiponectin. Adiponectin also interacts with a putative receptor, T-cadherin [30]; however, since T-cadherin lacks the transmembrane and cytoplasmic domains, the biological significance of this non-transmembrane receptor is not clear yet [31], T-cadherin receptor probably acts as a coreceptor to exert its functions with the integrin pathway or AdipoR1/R2 receptors [32].

It has been demonstrated that knockout of AdipoR1 gene in mice causes the abrogation of adiponectin-induced AMPK activation, whereas disruption of AdipoR2 gene in mice results in decreased activity of PPAR-α signaling pathways [33]. Furthermore, simultaneous disruption of both AdipoR1 and R2 has abolished adiponectin binding and actions, resulting in increased tissue triglyceride content, inflammation and oxidative stress in mice, and thus leading to insulin resistance and marked glucose intolerance. Therefore, AdipoR1 and R2 serve as the predominant receptors for adiponectin *in vivo* and play important roles in the regulation of glucose and lipid metabolism, inflammation and oxidative stress *in vivo*.

Thus, all of these above studies indicate that adiponectin functions as an important anti-inflammatory, antidiabetic and antiatherogenic protein in metabolism.

MACROPHAGE AND ITS ROLES

Macrophages are a heterogeneous population of phagocytic cells found throughout the body that originate from the mononuclear phagocytic system [34]. These are highly plastic cells that arise from circulating myeloid-derived blood monocytes that have entered target tissues and gained the phenotypic and functional attributes of their tissue of residence. Recent attention has focused on the potential role of macrophages in the process of metabolic diseases [35, 36]. It has been shown that in obesity, adipose tissue contains an increased number of resident macrophages and that in some circumstances, macrophages can constitute up to 40% of the cell population within an adipose tissue depot [37, 38]; and macrophages are obviously a potential source of secreted proinflammatory factors to other tissues for insulin resistance. This correlative evidence has led to the concept that macrophages can directly influence other insulin target tissues, such as skeletal muscle, liver, and adipose tissue, and cause whole body systemic insulin resistance. Furthermore, animal models have also been reported to evidence the causal role of the macrophage in leading to insulin resistance [39]. When these animals were fed with a high-fat diet, the macrophage-specific inflammatory pathway knockout mice were relatively protected from glucose intolerance and hyperinsulinemia, and these results showed a global improvement in insulin sensitivity in all insulin-target tissues. These studies are consistent with the interpretation that the macrophage is an important, and potentially initiating, cell type in the process of inflammation induced insulin resistance [40]. To prevent macrophage infiltration into these insulin-target tissues will have beneficial effects on the inflammatory response and the abnormal metabolic state.

Except of the effects on insulin sensitivity, macrophages also have critical impacts on cardiovascular disease in metabolic syndrome, such as atherosclerosis. Atherosclerosis is considered a chronic inflammatory disease and a disorder of lipid metabolism [41]. The complex physiopathologic process is initiated by the formation of cholesterol-rich lesions in the arterial wall. The accumulation of cholesterol-rich lipoproteins in the vascular artery wall results in the recruitment of circulating monocytes, their adhesion to the endothelium, and their differentiation into macrophages. Macrophages play a crucial role in this process because they accumulate large amounts of lipid to form the foam cells that initiate the formation of the lesion and participate actively in the development of the atherosclerotic lesion. Because the transformation of macrophages into foam cells is a critical component of atherosclerotic lesion formation [42], the prevention or reversal of cholesterol accumulation or the production of inflammatory mediators in macrophage foam cells could result in protection from multiple pathological effects of atherosclerosis and other abnormal metabolic disorders in metabolic syndrome.

As we know that adiponectin is almost exclusively expressed and secreted from adipocytes but with trace levels in other type of cells/tissues [43, 44], however, adiponectin expression has not been detected in macrophage cells. Since two adiponectin receptors, the AdipoR1 and AdipoR2, have been found to be expressed in human macrophages [45] with AdipoR1 predominating in these macrophage cells [46], therefore, it is generally considered that the binding of adiponectin to its receptors, and subsequent signaling events initiated by this binding, are probably responsible for the physiological effects in macrophage cells.

Human THP-1 monocytic cell line [47] is a well-characterized cell model for studying processes and mechanisms of monocyte differentiation into macrophage, and macrophage transformation into foam cell. These THP-1 monocytic cells can be induced to differentiate into macrophages following treatment with phorbol myristate acetate (PMA), and the resulting macrophages can then be induced to form foam cells following treatment with modified low-density lipoprotein (LDL) [48]. Adiponectin has been reported to inhibit lipid accumulation by down regulating scavenger receptor A (SR-A) expression and acyl-CoA: cholesterol-acyltransferase 1 (ACAT1) expression, two key players in foam cell formation, in cultured human monocyte-derived macrophages when the adipocyte-derived adiponectin protein was added to the THP-1 cell cultures [49, 50]. These results suggest adiponectin's inhibitory role in macrophage foam cell formation from human monocyte-derived macrophages. Interestingly, when the adiponectin gene was directly expressed in THP-1 cells, these monocyte-derived macrophage foam cells can significantly decrease triglyceride and cholesterol accumulation in these cells by reducing oxidized low-density lipoprotein (oxLDL) uptake into the cells while enhancing HDL-mediated cholesterol efflux [51]. There are two potential mechanisms for the reduced lipid accumulation in these adiponectin-transduced macrophage foam cells: the first mechanism involves the peroxisome proliferator-activated receptor gamma (PPAR-γ) and liver X receptor (LXR) signaling pathways which up-regulate the expression of ATP binding cassette A1 gene (ABCA1) and promote lipid efflux from these cells; the second mechanism involves decreased lipid uptake and increased lipid hydrolysis which may result from decreased scavenger receptor AI (SR-AI) and increased scavenger receptor BI (SR-BI) and hormone sensitive lipase (HSL) gene activities in the adiponectin-transduced macrophage foam cells. Furthermore, two inflammatory cytokines, monocyte chemoattractant protein 1 (MCP-1) and TNF-α, are also decreased in the adiponectin-transduced macrophage foam cells. These studies suggest that expression of adiponectin in human THP-1 macrophage cells can modulate multiple signaling pathways for both lipid metabolism and inflammatory response to reduce macrophage foam cell formation during atherosclerosis.

Macrophages can be functionally divided into two subsets with two different phenotypes, classically activated M1 and alternatively activated M2 macrophages, based on these cells in response to exposure to microenvironmental stimuli and signals [52, 53]. M1 and M2 macrophages in adipose tissue can be distinguished by the presence or the absence of CD11c, an M1 macrophage marker [54]. M1 macrophages produce high levels of proinflammatory cytokines, such as TNF-α, MCP-1 and IL-6, and play a role in tissue destruction. In contrast, M2 are more prominent producers of anti-inflammatory cytokines, IL-1 receptor antagonist (IL-1Ra), IL-10 and transforming growth factor beta (TGF-β), which are involved in tissue repair or remodeling. It has been reported that adiponectin can modulate macrophage polarization from a classically activated M1 phenotype to an alternatively-activated M2 phenotype in mouse adipose tissue. These studies have shown that peritoneal macrophages and the stromal vascular fraction (SVF) cells of adipose tissue isolated from adiponectin knock-out mice display increased M1 markers and decreased M2 markers, and the systemic delivery of adenovirus expressing adiponectin significantly augmented M1 markers in peritoneal macrophages and SVF cells in both wild-type and adiponectin knock-out mice [55]. Adiponectin also stimulates the expression of M2 markers and attenuated the expression of M1 markers in human monocyte-derived macrophages and SVF cells isolated from human adipose tissue. These studies have demonstrated that adiponectin functions as a regulator of macrophage polarization, and they indicate that conditions of high adiponectin expression may deter metabolic and cardiovascular disease progression by favoring an anti-inflammatory phenotype in macrophages.

Recent studies using adiponectin transgenic mice (Ad-TG) that specifically express adiponectin in mouse macrophages have further demonstrated that adiponectin can physiologically modulate metabolic activities in vivo by improving metabolism in distal tissues for metabolic syndrome [56, 57]. In these studies, an adiponectin macrophage-specific transgenic mouse model has been generated using a scavenger receptor A-I gene enhancer/promoter; data from these adiponectin transgenic mice have shown that expression of adiponectin in mouse macrophages can significantly reduce cholesterol and triglyceride accumulation in these mouse macrophages and bloodstream with significantly lower levels of total cholesterol due to significantly decreased LDL and VLDL (very-low-density lipoprotein) cholesterol concentrations together with a significant increase in HDL cholesterol, which contribute to the reduction of macrophage foam cell formation in the atherosclerotic lesions when these macrophage modified transgenic animals are crossed with the LDL receptor (Ldlr)-deficient mouse model under a high-fat diet condition. In addition, Ad-TG mice also exhibit higher total and high-molecular weight adiponectin levels in plasma. Furthermore, the adiponectin-modified mouse macrophages also alter systemic metabolism in other metabolic active tissues, such as skeletal muscle, liver and adipose tissue, with cell interactions either through infiltration or circulation by bloodstream (Figure 1), which influences the levels of inflammatory cytokines, MCP-1, TNF-α, and IL-6 in the metabolic active tissues and thus physiologically improves whole-body glucose tolerance and insulin sensitivity. These studies indicate that macrophages engineered to produce adiponectin can significantly improve intercellular microenvironment and alter the important metabolic activities in other tissues through macrophage infiltration and circulation to reduce inflammation response and lipid accumulation in these tissues. Although adiponectin is a specific adipokine expressed and secreted physiologically mainly from the adipocytes, these studies have demonstrated that adiponectin from the modified macrophages can also exert powerful anti-inflammatory and antiatherogenic effects on whole-body metabolism. All of these results further support studies addressing the use of genetically manipulated macrophages as a novel therapeutic approach for treatment of metabolic diseases.

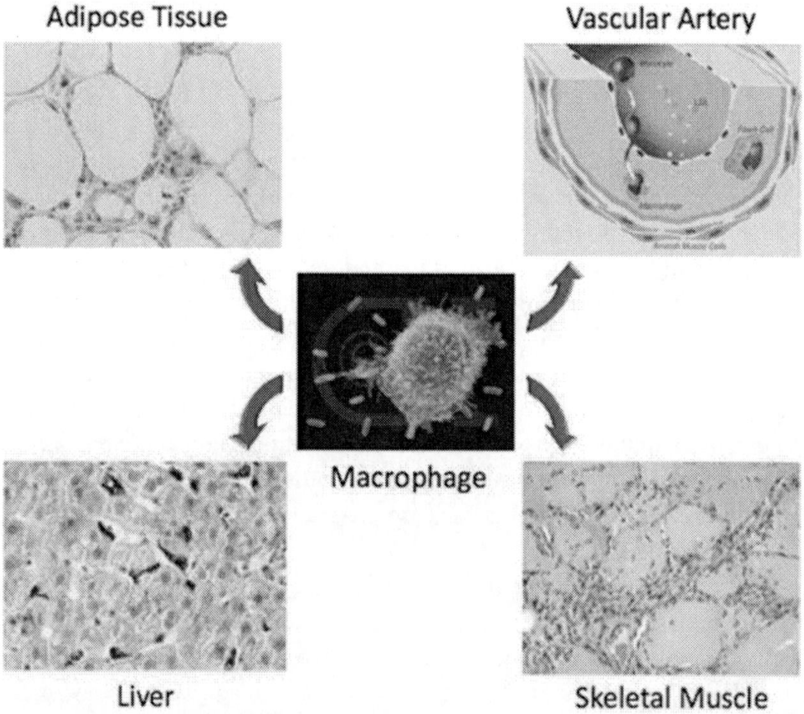

Figure 1. Adiponectin modified macrophages to influence other metabolic tissues. Adiponectin is expressed in macrophages and the modified macrophages infiltrate/circulate or reside in metabolic active tissues such as adipose tissue, liver, vascular artery, and skeletal muscle through blood vessels. The modified macrophages interact with these metabolic tissues/cells to improve intercellular microenvironment for whole body systemic metabolism. (Source: Our unpublished data for adipose tissue; adapted from *Netterimages.com* for liver; *Metabolic Alchemy Forum* for macrophage; *Lens* magazine for vascular artery; and *Veterinary Pathology Forum* for skeletal muscle.

CONCLUSION

Metabolic syndrome is a combination of several serious metabolic disorders, including obesity, insulin resistance, type II diabetes, and cardiovascular disease. Currently, there is no effective way to treat all the components in metabolic syndrome except by lifestyle changes. Adiponectin has been proved as one adipokine with antidiabetes, anti-inflammatory, and antiatherogenic properties mainly expressed and secreted from adipose tissues. Scientific studies conducted in human and animal models of diabetes, obesity, and atherosclerosis have clearly demonstrated that adiponectin can indeed have beneficial effects on these diseases. The use of macrophages as carriers for adiponectin will provide a novel and unique strategy for studying the mechanisms of adiponectin-mediated alterations in body metabolism in vivo. This unique strategy can probably serve as a potential novel and innovative therapeutic approach for treatment of the metabolic syndrome in the future.

REFERENCES

[1] G. M. Reaven, *Diabetes* 37, 1595 (1988).
[2] G. M. Reaven, *Diabetes Care* 14, 195 (1991).
[3] R. A. DeFronzo, *Diabetes Care* 14, 173 (1991).
[4] R. W. Nesto, D. Bell, R. O. Bonow, V. Fonseca, S. M. Grundy, E. S. Horton, M. L. Winter, D. Porte, C. F. Semenkovich, S. Smith, L. H. Young and R. Kahn, *Circulation* 108, 2941 (2003).
[5] S. E. Nissen and K. Wolski, *N. Engl. J. Med.* 356, 2457 (2007).
[6] J. D. Lewis, A. Ferrara, T. Peng, M. Hedderson, W. B. Bilker, C. P. Quesenberry Jr., D. J. Vaughn, L. Nessel, J. Selby, B. L. Strom, *Diabetes Care* 34, 923 (2011).
[7] Y. Arita, S. Kihara, N. Ouchi, M. Takahashi, K. Maeda, J. Miyagawa, K. Hotta, I. Shimomura, T. Nakamura, K. Miyaoka, H. Kuriyama, M. Nishida, S. Yamashita, K. Okubo, K. Matsubara, M. Muraguchi, Y. Ohmoto, T. Funahashi and Y. Matsuzawa, *Biochem. Biophys. Res. Commun.* 257, 79 (1999).
[8] K. Maeda, K. Okubo, I. Shimomura, T. Funahashi, Y. Matsuzawa and K. Matsubara, *Biochem. Biophys. Res. Commun.* 221, 286 (1996).
[9] E. Hu, P. Liang and B. M. Spiegelman, *J. Biol. Chem.* 271, 10697 (1996).
[10] Y. Nakano, T. Tobe, N. H. Choi-Miura, T. Mazda and M. Tomita, *J. Biochem.* 120, 803 (1996).
[11] P. E. Scherer, S. Williams, M. Fogliano, G. Baldini and H. F. Lodish, *J. Biol. Chem.* 270, 26746 (1995).
[12] H. Berg, T. P. Combs and P. E. Scherer, *Trends Endocrinol. Metab.* 13, 84 (2002).
[13] U. B. Pajvani, M. Hawkins, T. P. Combs, M. W. Rajala, T. Doebber, J. P. Berger, J. A. Wagner, M. Wu, A. Knopps, A. H. Xiang, K. M. Utzschneider, S. E. Kahn, J. M. Olefsky, T. A. Buchanan and P. E. Scherer, *J. Biol. Chem.* 279, 12152 (2004).
[14] C. Lara-Castro, Y. Fu, B. H. Chung and W. T. Garvey, *Curr. Opinion Lipidology* 18, 263 (2007).
[15] K. Hotta, T. Funahashi, Y. Arita, M. Takahashi, M. Matsuda, Y. Okamoto, H. Iwahashi, H. Kuriyama, N. Ouchi, K. Maeda, M. Nishida, S. Kihara, N. Sakai, T. Nakajima, K.

Hasegawa, M. Muraguchi, Y. Ohmoto, T. Nakamura, S. Yamashita, T. Hanafusa and Y. Matsuzawa, *Arterioscler. Thromb. Vasc. Biol.* 20, 1595 (2000).
[16] R. S. Lindsay, T. Funahashi, R. L. Hanson, Y. Matsuzawa, S. Tanaka, P. A. Tataranni, W. C. Knowler and J. Krakoff, *Lancet* 360, 57 (2002).
[17] R. S. Lindsay, T. Funahashi, J. Krakoff, Y. Matsuzawa, S. Tanaka, S. Kobes, P. H. Bennett, P. A. Tataranni, W. C. Knowler, R. L. Hanson, *Diabetes* 52, 2419 (2003).
[18] K. Hotta, T. Funahashi, N. L. Bodkin, H. K. Ortmeyer, Y. Arita, B. C. Hansen and Y. Matsuzawa, *Diabetes* 50, 1126 (2001).
[19] C. Weyer, T. Funahashi, S. Tanaka, K. Hotta, Y. Matsuzawa, R. E. Pratley and P. A. Tataranni, *J. Clin. Endocrinol. Metab.* 86, 1930 (2001).
[20] C. Menzaghi, V. Trischitta and A. Doria, *Diabetes* 56, 1198 (2007).
[21] H. Green and O. Kehinde, *Cell* 5, 19 (1975).
[22] Y. Fu, N. Luo, R. L. Klein and W. T. Garvey, *J. Lipid Res.* 46, 1369 (2005).
[23] T. P. Combs, U. B. Pajvani, A. H. Berg, Y. Lin, L. A. Jelicks, M. Laplante, A. R. Nawrocki, M. W. Rajala, A. F. Parlow, L. Cheeseboro, Y. Y. Ding, R. G. Russell, D. Lindemann, A. Hartley, G. R. Baker, S. Obici, Y. Deshaies, M. Ludgate, L. Rossetti and P. E. Scherer, *Endocrinology* 145, 367 (2004).
[24] J. Y. Kim, E. van de Wall, M. Laplante, A. Azzara, M. E. Trujillo, S. M. Hofmann, T. Schraw, J. L. Durand, H. Li, G. Li, L. A. Jelicks, M. F. Mehler, D. Y. Hui, Y. Deshaies, G. I. Shulman, G. J. Schwartz and P. E. Scherer, *J. Clin. Invest.*, 117, 2621 (2007).
[25] W. Asterholm and P. E. Scherer, *Am. J. Pathol.* 176, 1364 (2010).
[26] N. Kubota, Y. Terauchi, T. Yamauchi, T. Kubota, M. Moroi, J. Matsui, K. Eto, T. Yamashita, J. Kamon, H. Satoh, W. Yano, P. Froguel, R. Nagai, S. Kimura, T. Kadowaki, T. Noda, *J. Biol. Chem.* 277, 25863 (2002).
[27] N. Maeda, I. Shimomura, K. Kishida, H. Nishizawa, M. Matsuda, H. Nagaretani, N. Furuyama, H. Kondo, M. Takahashi, Y. Arita, R. Komuro, N. Ouchi, S. Kihara, Y. Tochino, K. Okutomi, M. Horie, S. Takeda, T. Aoyama, T. Funahashi, Y. Matsuzawa, *Nat. Med.* 8, 731 (2002).
[28] N. Ouchi, M. Ohishi, S. Kihara, T. Funahashi, T. Nakamura, H. Nagaretani, M. Kumada, K. Ohashi, Y. Okamoto, H. Nishizawa, K. Kishida, N. Maeda, A. Nagasawa, H. Kobayashi, H. Hiraoka, N. Komai, M. Kaibe, H. Rakugi, T. Ogihara, Y. Matsuzawa, *Hypertension* 42, 231 (2003).
[29] T. Yamauchi, J. Kamon, Y. Ito, A. Tsuchida, T. Yokomizo, S. Kita, T. Sugiyama, M. Miyagishi, K. Hara, M. Tsunoda, K. Murakami, T. Ohteki, S. Uchida, S. Takekawa, H. Waki, N. H. Tsuno, Y. Shibata, Y. Terauchi, P. Froguel, K. Tobe, S. Koyasu, K. Taira, T. Kitamura, T. Shimizu, R. Nagai and T. Kadowaki, *Nature* 423, 762 (2003).
[30] C. Hug, J. Wang, N. S. Ahmad, J. S. Bogan, T. S. Tsao and H. F. Lodish, *Proc. Natl. Acad. Sci. U. S. A.* 101, 10308 (2004).
[31] T. Kadowaki, T. Yamauchi, N. Kubota, K. Hara, K. Ueki and K. Tobeet, *J. Clin. Invest.* 116, 1784 (2006).
[32] M. S. Denzel, M-C. Scimia, P. M. Zumstein, K. Walsh, P. Ruiz-Lozano and B. Ranscht, *J. Clin. Invest.* 120, 4342 2010.
[33] T. Yamauchi, Y. Nio, T. Maki, M. Kobayashi, T. Takazawa, M. Iwabu, M. Okada-Iwabu, S. Kawamoto, N. Kubota, T. Kubota, Y. Ito, J. Kamon, A. Tsuchida, K. Kumagai, H. Kozono, Y. Hada, H. Ogata, K. Tokuyama, M. Tsunoda, T. Ide, K.

Murakami, M. Awazawa, I. Takamoto, P. Froguel, K. Hara, K. Tobe, R. Nagai, K. Ueki and T. Kadowaki, *Nat. Med.* 13, 332 (2007).
[34] Mantovani, A. Sica and M. Locati, *Immunity* 23, 344 (2005).
[35] Bouloumie, C. A. Curat, C. Sengenes, K. Lolmede, A. Miranville and R. Busse, *Curr. Opin. Clin. Nutr. Metab. Care* 8, 347 (2005).
[36] H. Kanda, S. Tateya, Y. Tamori, K. Kotani, K. Hiasa, R. Kitazawa, S. Kitazawa, H. Miyachi, S. Maeda, K. Egashira and M. Kasuga, *J. Clin. Invest.* 116, 1494 (2006).
[37] S. P. Weisberg, D. McCann, M. Desai, M. Rosenbaum, R. L. Leibel and A. W. Ferrante, *J. Clin. Invest.* 112, 1796 (2003).
[38] H. Xu, G. T. Barnes, Q. Yang, G. Tan, D. Yang, C. J. Chou, J. Sole, A. Nichols, J. S. Ross, L. A. Tartaglia and H. Chen, *J. Clin. Invest.* 112, 1821 (2003).
[39] M. C. Arkan, A. L. Hevener, F. R. Greten, S. Maeda, Z-W. Li, J. M. Long, A. Wynshaw-Boris, G. Poli, J. Olefsky and M. Karin, *Nat. Med.* 11, 191 (2005).
[40] M. T. A. Nguyen, S. Favelyukis, A. K. Nguyen, D. Reichart, P. A. Scott, A. Jenn, R. Liu-Bryan, C. K. Glass, J. G. Neels and J. M. Olefsky, *J. Biol. Chem.* 282, 35279 (2007).
[41] Lusis, *Nature* 407, 233 (2000).
[42] R. Ross, *Nature* 362, 801 (1993).
[43] S. Maddineni, S. Metzger, O. Ocon, G. Hendricks and R. Ramachandran, *Endocrinology* 146, 4250 (2005).
[44] 4. M. Nishida, T. Funahashi and I. Shimomura, *Med. Mol. Morphol.* 40, 55 (2007).
[45] G. Chinetti, C. Zawadski, J. C. Fruchart and B. Staels, *Biochem. Biophys. Res. Commun.* 314, 151 (2004).
[46] N. Yamaguchia, J. G. Argueta, Y. Masuhiro, M. Kagishita, K. Nonaka, T. Saito, S. Hanazawa and Y. Yamashita, *FEBS Letters* 579, 6821 (2005).
[47] Auwerx, *Experientia* 47, 22 (1991).
[48] D. Steinberg, *J. Biol. Chem.* 272, 20963 (1997).
[49] N. Ouchi, S. Kihara, Y. Arita, M. Nishida, A. Matsuyama, Y. Okamoto, M. Ishigami, H. Kuriyama, K. Kishida, H. Nishizawa, K. Hotta, M. Muraguchi, Y. Ohmoto, S. Yamashita, T. Funahashi and Y. Matsuzawa, *Circulation* 103, 1057 (2001).
[50] Furukawa, M. Hori, N. Ouchi, S. Kihara, T. Funahashi, Y. Matsuzawa, A. Miyazaki, H. Nakayama and S. Horiuchi, *Biochem. Biophys. Res. Commun.* 317, 831 (2004).
[51] Tian, N. Luo, R. L. Klein, B. H. Chung, W. T. Garvey and Y. Fu, *Atherosclerosis* 202, 152 (2009).
[52] S. Gordon, *Nat. Rev. Immunol.* 3, 23 (2003).
[53] Mantovani, A. Sica, S. Sozzani, P. Allavena, A. Vecchi and M. Locati, *Trends Immunol.* 25, 677 (2004).
[54] N. Lumeng, J. L. Bodzin and A. R. Saltiel, *J. Clin. Invest.* 117, 175 (2007).
[55] K. Ohashi, J. L. Parker, N. Ouchi, A. Higuchi, J. A. Vita, N. Gokce, A. A. Pedersen, C. Kalthoff, S. Tullin, A. Sams, R. Summer and K. Walsh, *J. Biol. Chem.* 285, 6153 (2010).
[56] Luo, J. Liu, B. H. Chung, Q. Yang, R. L. Klein, W. T. Garvey and Y. Fu, *Diabetes* 59,791 (2010).
[57] Luo, X. Wang, B. H. Chung, M. H. Lee, R. L. Klein, W. T. Garvey and Y. Fu, *Am. J. Physiol. Endocrinol. Metab.* Apr 19 (2011).

In: Adiponectin: Production, Regulation and Roles in Disease ISBN: 978-1-61942-165-3
Editor: Masayoshi Yamaguchi

Chapter III

ADIPONECTIN, ITS CARDIOPROTECTIVE EFFECTS AND GENETIC VARIABILITY IN RELATION TO RESTENOSIS AFTER PERCUTANEOUS CORONARY INTERVENTION (PCI)

J. Bienertová-Vašků*, P. Bienert, and A. Vašků
Department of Pathological Physiology, Faculty of Medicine,
Masaryk University Brno, Kamenice Brno, Czech Republic

ABSTRACT

Restenosis after percutaneous coronary intervention (PCI) represents a major limitation to this procedure and despite large efforts it is still difficult to stratify patients after PCI according to restenosis risk. A wide range of proteins have been investigated in order to establish new diagnostic/prognostic markers of restenosis after PCI, without major effects.

Adiponectin, a protein secreted exclusively by the adipose tissue, has recently emerged as an anti-inflammatory, anti-atherogenic, insulin-sensitizing factor providing possible link between white adipose tissue and cardiovascular disease. This crucial role in cardioprotection is mediated via a variety of mechanisms, including inhibition of foam cell formation, expression of antithrombogenic properties and inhibition of smooth muscle cells proliferation, all vital processes in restenosis development.

Considering its cardioprotective effects, the role of adiponectin in prediction of adverse cardiovascular events has been studied extensively. However, little attention is paid to adiponectin role in restenosis development after PCI, even though pathophysiology of restenosis covers directly the mechanisms regulated or substantially influenced by adiponectin. The current review describes pathophysiological effects of adiponectin in cardiovascular system, mainly those closely associated with restenosis development and attention is being paid also to the role of genetic variability.

Keywords: adiponectin; percutaneous coronary intervention; restenosis; polymorphism

* Corresponding author: Bienertová-Vašků, J., Department of Pathological Physiology, Faculty of Medicine, Masaryk University Brno, Kamenice 5 A18, 625 00 Brno, Czech Republic, email: jbienert@med.muni.cz, tel.: +420-739-300-485, fax: +420-549-494-340.

Introduction

For decades, white adipose tissue (WAT) has been recognized as an inactive deposit of triglycerides (TG), releasing free fatty acids/glycerol in response to varying energy demands. However, this somewhat mechanistic perception of the physiological role of WAT has undergone a complex and fundamental change after discovery of leptin (OMIM 164160) in 1994 [1], which demonstrated that adipose tissue represents an extremely active endocrine organ secreting numerous mediators, commonly known as adipokines.

Adiponectin (OMIM 605441) was originally identified by four independent research groups in the 1990s using different experimental approaches, in both mice and humans [2-5], and since its discovery, it attracted much attention particularly in the field of molecular pathophysiology of insulin resistance and type 2 diabetes mellitus (T2D). Not long after its discovery, the potential of adiponectin as an anti-atherogenic factor was recognized due to its ability to regulate expression of a variety of endothelial adhesion molecules [6]. Furthermore, adiponectin differs from other adipokines in being secreted exclusively by the adipose tissue, being present in abundant amount in the circulation, and found at reduced levels in the presence of obesity. The numerous experimental as well as clinical studies that were extensively carried out in order to investigate the role of adiponectin in physiological regulation of blood vessels yielded rather contradictory results, making the virtual physiological and pathophysiological role of adiponectin in cardiovascular disease rather elusive.

This chapter summarizes genetic and pathophysiological relevancies of adiponectin and discusses its value as a biomarker in restenosis diagnosis/prediction, both as a plasma protein and as the genetic single nucleotide polymorphisms (SNPs) in the adiponectin locus.

Structural and Positional Characteristics of Adiponectin

Genome wide scan studies indicate that a susceptibility locus linked to T2D, obesity and coronary heart disease may be linked to the 3q27 chromosomal region [7], where the adiponectin gene is located. The adiponectin gene, ADIPOQ, (alternatively ACDC, APM1, ACRP30 or GBF28) spans 1.579 kb and contains 3 exons. ADIPOQ gene is generally considered an important candidate for T2D, obesity or diabetic nefropathy.

Adiponectin encodes a secretory protein that structurally belongs to the complement 1q family. Full-length adiponectin consists of 247 amino acid residues including the N-terminal hypervariable region, followed by a conserved collagenous domain comprising 22 Gly-Xaa-Yaa repeats and a C-terminal C1q-like globular domain. In plasma, adiponectin combines via its globular or collagen domains to form various multimer complexes. The three major complexes in plasma are a low-molecular-weight trimer (via globular domain interactions), a middle-molecular-weight hexamer, and a high-molecular-weight 12- to 18-mer (via collagenous domain interactions) [8]. Adiponectin is known to be abundantly present in the circulation, accounting for approx. 0.01% of the total human plasma protein [9]. Unlike most other adipokines, circulating levels of adiponectin are decreased significantly in obesity and its related co-morbidities, mainly T2D, insulin resistance, coronary heart disease, stroke, non-alcoholic steatohepatitis, and cancer (breast, colon and prostate) [10-12].

Adiponectin action is mediated via its two receptors, designated AdipoR1 and AdipoR2, which are primarily expressed in skeletal muscle and liver, respectively [13] and have seven transmembrane domains differing structurally from G protein-coupled receptors and serving to bind full-length and globular isoform of adiponectin. AdipoR1 was subsequently demonstrated to be present in endothelial cells [14] and cardiomyocytes [15], whereas AdipoR2 is expressed also on endothelial cells. These findings have important implications as they provide a potential molecular pathway of adiponectin effects both in endothelium and in myocardium, involving activation of the AMP-activated protein kinase (AMPK), the peroxisome proliferator-activated receptor (PPAR)-γ, and the p38 mitogen-activated protein kinase (MAPK)-signaling pathways [16].

Cardioprotective Effects of Adiponectin

The initial direct evidence for adiponectin's potential role in cardioprotection was provided by animal experiments. Consistent with the supposed vascular-protective role of adiponectin, adiponectin-deficient mice have been shown to be more susceptible to develop vascular disorders.

Adiponectin-deficient or knockout mice, when compared to wild-type controls, show neointimal thickening and enhanced proliferation of vascular smooth muscle cells after mechanical injury to arteries [17]. Adiponectin-deficient mice also show low nitric oxide availability and up-regulation of endothelial adhesion molecules, both of which can be reversed by exogenous adiponectin administration [18].

Furthermore, adiponectin overexpression in apolipoprotein E deficient mice that are animal models of atherosclerosis prevents the conversion of fatty streaks to mature atherosclerotic plaques [19]. In a similar experimental design, apoE knockout mice overexpressing globular adiponectin presented with fewer atherosclerotic lesions when compared with control apoE knockout mice, despite similar plasma glucose and lipid levels [20].

Generally, hypoadiponectinemia seems to be associated with impaired endothelium-dependent vasorelaxation [21]. In this study on adiponectin knocked-out mice, the authors observed that the endothelium-dependent vasodilation in response to acetylcholine was significantly reduced in adiponectin-null mice compared with *wt* mice.

In addition to its insulin-sensitising effects, several mechanisms of cardioprotective action of adiponectin have been recently recognized (Figure 1).

Adiponectin has been suggested to significantly affect basic inflammatory mechanisms involved in molecular pathophysiology of cardiovascular disease, mainly with respect to atherogenesis. However, these mechanisms are also inseparably linked to development of restenosis and understanding of the general cardioprotective effects of adiponectin provides a good basis for further insight into the role of adiponectin in restenosis development.

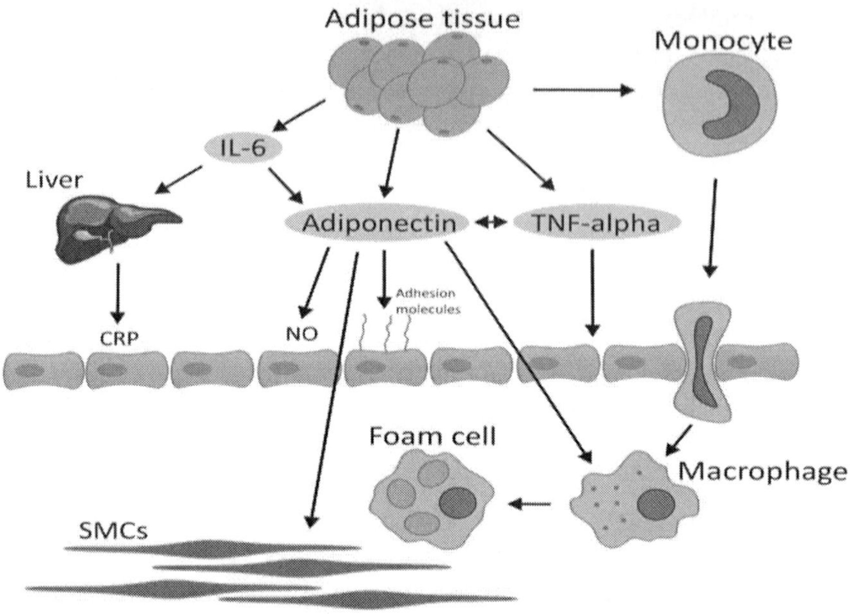

Figure 1. Vascular effects of adiponectin at the level of vascular wall.

The Vascular Effects of Adiponectin

The crucial report by Ouchi et al [22] demonstrates interference of adiponectin with nuclear factor kappa-B (NF-kB) intracellular signaling pathways via cAMP dependent pathway which has important implications mainly with respect to TNF-alfa, a factor initiating and regulating inflammatory processes in vascular tissue, whose expression is tightly controlled by NF-kB. Moreover, adiponectin and TNF-α reported to reciprocally inhibit each other's production in adipose tissue [23]. Adiponectin induces expression of a variety of anti-inflammatory mediators, such as IL-1RA or IL-10 [24-25]. In cultured endothelial cells, adiponectin attenuated HB-EGF expression stimulated by tumor necrosis factor alpha. This study proposes an existence of adipo-vascular axis, a direct link between fat and arteries [17]. Adiponectin gene expression as well as secretion is further inhibited by IL6 in 3T3-L1 adipocytes [26].

Antithrombotic Effects of Adiponectin

Adiponectin-deficient mice experience accelerated thrombus formation following laser-induced carotid arterial injury, whereas adenovirus-mediated expression of adiponectin reverses these changes [27]. These results suggest adiponectin plays a role in prevention of thrombogenesis. In vitro, adiponectin inhibits collagen-induced platelet aggregation in platelets obtained from human subjects and adiponectin-deficient mice [27]. It has been also suggested that adiponectin mediates the suppressive effect of rosiglitazone on plasminogen activator inhibitor-1 production [28].

Enhancement of Endothelial NO Production

In *in vitro* studies, adiponectin was reported to directly affect the cellular components of the vascular wall, to activate endothelial nitric oxide synthase (eNOS) via PI3-dependent pathways and thus to promote release of nitric oxide (NO) from endothelial cells [29-30]. The activation and subsequent up-regulation of eNOS could be partially responsible for some of the reported vaso-protective properties of adiponectin.

Antioxidant Activities of Adiponectin in the Endothelium

The vasoprotective action of adiponectin could be further explained by reducing NADPH oxidase-derived superoxide generation which might affect the redox state of the endothelial cells [31]. In addition, the cardioprotective effects of adiponectin against myocardial ischaemia/reperfusion might also be attributed, at least in part, to the reduction of oxidative stress by adiponectin [32].

Effects of Adiponectin on Expression of Endothelial Adhesion Molecules

At the level of vascular endothelium, adiponectin is capable of suppressing the expression of cell adhesion molecules [e.g., intercellular adhesion molecule (ICAM)-1, vascular cell adhesion molecule (VCAM)-1, and E-selectin] which can further result in reduced adhesion of monocytes to the vascular wall [22], which has potentially many important implications for development of a variety of adverse cardiovascular events. By contrast, a more recent study has demonstrated that acute treatment with globular adiponectin activates NF-kB and enhances the expression of adhesion molecules and monocyte chemotactic protein-1 (MCP-1) in endothelial cells, through activation of the sphingosine kinase signalling pathway [33]. The discrepancy between these reports can be attributed to different isoforms of adiponectin that were investigated or to different laboratory methodology used in these studies.

Inhibition of Macrophage Activation and Prevention of Foam Cell Formation

Infiltration of monocytes into the subendothelial space is a critical event in endothelial dysfunction, with huge implications in restenosis development. Adiponectin harbours strong potential of preventing lipid accumulation and foam cells formation by suppressing the expression of scavenger receptors on macrophages [34].

Adiponectin also has inhibitory effects on the proliferation of myelomonocytic lineage cells and on the function of matured macrophages [35].

Suppressing Proliferation and Migration of Smooth Muscle Cells

Adiponectin inhibits both proliferation and migration of human aortic SMCs induced by several atherogenic growth factors, including HB-EGF [heparinbinding EGF (epidermal growth factor)-like growth factor], PDGF-BB (platelet-derived growth factor-BB) and bFGF (basic fibroblast growth factor) [9]. Adiponectin exerts its effect by directly binding PDGF-BB and inhibiting growth-factor-stimulated extracellular signal-regulated kinase signaling. Therefore, it might be suggested adiponectin plays a significant role in vascular remodelling as well.

POTENTIAL OF ADIPONECTIN PLASMA LEVEL AS A BIOMARKER FOR VARIOUS CARDIOVASCULAR CONDITIONS

Atherosclerosis Progression/Regression

In the Salzburg Atherosclerosis Prevention Program (SAPHIR) [36] study comprising a total of 1515 middle-aged healthy white subjects, the authors investigated the common carotid artery intima-media thickness (CIMT) and presence of atherosclerotic plaques by B-mode ultrasound. Based on the outcomes of the study, it was suggested that adiponectin levels are negatively associated with CIMT, concluding hypoadiponectinemia might represent a risk factor for development of early atherosclerosis. In a recent study by Dullaart RP [37], adiponectin plasma level was negatively correlated with carotid intimal thickness, independently of leptin plasma levels, in a cohort of 161 middle-aged men and women. However, Shimetani et al [38] did not observe any significant relationships between intima-media complex thickness and circulating adiponectin levels in their cohort of Japanese men and women. Nevertheless, the circulating adiponectin level seems to be a useful diagnostic marker of atherosclerosis progression, as most of the studies confirm a strong negative correlation between the carotid intima thickness and adiponectin plasma/serum levels.

Peripheral Arterial Occlusive Disease (PAOD)

The reports on significance of adiponectin levels in PAOD are controversial. In the study by Iwashima Y et al [39], the plasma levels of adiponectin as well as markers of inflammation and endothelial function in peripheral arterial occlusive disease (PAOD) were investigated in a cohort of 88 subjects.

The authors observed a tendency to lower adiponectin levels after percutaneous transluminal angioplasty, although Hs-CRP levels significantly increased. Moreover, plasma levels of adiponectin in these PAOD patients were reduced in proportion to the severity of the disease. However, in another study by Golledge J [40], no association of PAOD with circulating levels of adiponectin was observed and further studies are necessary to investigate the role of adiponectin in PAOD.

Saphenous Vein Graft Disease Post Coronary Artery Bypass Graft Surgery (CABG)

In the study by Kiris et al [41], the serum concentrations of adiponectin and CIMT were measured in 84 consecutive patients who underwent CABG. The results of the study indicate the presence of an inverse relationship between serum concentrations of adiponectin and subclinical carotid atherosclerosis in patients undergoing CABG. However, the potential of adiponectin levels for prediction of post-surgery saphenous vein graft disease was not investigated in this study. In another study by Eyileten Z [42] on CABG patients, that presented generally with low plasma adiponectin levels, plasma adiponectin level was increased one month after CABG, although the level was still lower than the control value. Based on the data from available studies, adiponectin might be a useful marker of progression risk after CABG.

Restenosis after Coronary PCI

Only a limited number of studies investigated the relationship between adiponectin plasma levels and restenosis risk after PCI, moreover, most of the studies were clearly underpowered for any significant conclusions.

In 2007 Shioji K [43] et al evaluated serum adiponectin levels in 184 patients undergoing PCI concluding that adiponectin plasma levels might be useful in prediction of adverse cardiovascular events after PCI. In the same year, Inami N et al [44] in their report suggest that restenosis development after PCI in patients with coronary artery disease (CAD) may involve the participation of MCP-1 after PCI, and that adiponectin incompletely prevent this MCP-1-dependent restenosis. However, their study comprised only 69 individuals and was apparently underpowered for any generalizations.

Kitta Y et [45] in their study from 2008 suggest that low adiponectin levels have a predictive value for late ISR after PCI with bare-metal stents in native coronary arteries. Based on the available data, adiponectin seems to be a good marker of ISR after PCI in CAD patients.

Coronary Heart Transplant Vasculopathy

Ambrosi P et al [46] investigated the circulating levels of adiponectin in a cohort of 95 heart recipients. In their cohort, plasma adiponectin was significantly lower in patients who presented also with metabolic syndrome, however, adiponectin levels were usually in the normal or high range and were associated lower creatinine clearance. Plasma adiponectin was not significantly different between patients with angiographic evidence of graft vasculopathy or without it. Replication of these data on larger population samples of different ethnicity seems to be crucial.

ADIPONECTIN AND CLINICAL DATA IN CARDIOLOGY

Clinical epidemiological studies associating plasma adiponectin with cardiovascular risks are controversial, the results of smaller populations-based studies being mostly questioned by following studies on larger population samples. This could be partially explained by the small statistical power of the initial studies that led to speculative conclusions, however, it might also unmask the complex pleiotropic effects of adiponectin both as a paracrine and an endocrine factor. The results of the large-scale population studies carried out to investigate possible relationship between adiponectin plasma levels and adverse cardiovascular events, are contradictory. In 2004, an extensive study [47] was performed to assess prospectively whether plasma adiponectin concentrations are associated with risk of myocardial infarction, in the total of 18 225 male participants of the Health Professionals Follow-up Study aged 40 to 75 years who were free of diagnosed cardiovascular disease at the time of blood draw (1993-1995). The authors concluded that high plasma concentrations of adiponectin were associated with lower risk of MI in men.

In 2006, Sattar et al [48] published a study summarizing measurements of baseline adiponectin levels in stored serum samples of 589 men with fatal CHD or nonfatal myocardial infarction and in 1231 controls nested within a prospective study of 5661 men. The authors also performed a meta-analysis of seven population-based prospective studies on involving a total of 1318 CHD cases, reporting the odds ratio of 0.84 [95% CI, 0.70 to 1.01] for men in the top third of adiponectin concentrations compared with those in the bottom third, which is similar to the results described in the experimental cohort: OR=0.89 (95% CI, 0.67 to 1.18). The authors conclude that any association with CHD risk is comparatively moderate and requires further investigation. However, in the British Women's Heart and Health Study [49], adiponectin did not predict coronary heart disease events both in adjusted or unadusted analyses. Other studies were carried out to investigate the relationship between adiponectin plasma levels and atherosclerosis risk. In addition, adiponectin was also reported to be associated with direct atheroma risk [50-51], which is in concordance with the current perception of atherosclerosis as an inflammatory disease. Generally, individuals at risk of atherosclerosis development express high levels of circulating inflammatory markers, including acute phase proteins, such as CRP [52]. Of interest, CRP levels were in clinical studies correlated inversely with adiponectin levels [53], further supporting the perception of adiponectin as an anti-inflammatory marker.

GENETIC VARIABILITY IN ADIPOQ LOCUS AND CARDIOVASCULAR DISEASE

Numerous single nucleotide polymorphisms in the adiponectin locus were associated with obesity, insulin resistance and/or T2D. However, only a very limited number of studies investigated utility of the common polymorphisms in the adiponectin / adiponectin receptor gene in diagnosis/prediction of adverse cardiovascular events.

In 2007, the total of 5 common single nucleotide polymorphisms (SNPs) at the adiponectin locus (rs266729, -11365C>G; rs822395, -4034A>C; rs822396, -3964A>G; rs2241766, +45T>G; and rs1501299, +276G>T) were investigated in the US Health

Professionals Follow-up Study (comprising 249 women and 266 men with non-fatal myocardial infarction) [54]. The study was focused on coronary heart disease, concluding that none of the investigated SNPs, including +45T/G, was significantly associated with development of CHD. However, even if the relationship between the SNPs and percutanoues coronary intervention was investigated in this study, no restenosis data were evaluated in this study.

In a study by Vasseur F [55] et al that was published in 2010, a total of 19 SNPs in 10 genes including ADIPOQ were genotyped in a cohort of 846 subjects. The authors used both the multifactor dimensionality reduction method to explain clusters of high and low LDL-C, and the restricted partition method to explain LDL-C levels. Both methods showed that BMI and haplotypes at the adiponectin encoding gene but not adiponectin level itself, were discriminant regarding to LDL-C. Subjects bearing an at-risk combination of BMI and ADIPOQ genotypes were prone to have a higher LDL-C (OR=3.13, 95% CI=2.20-4.46, $P<0.0001$) in this study. Also, no coronarographic data were provided in this study.

Common Polymorphism +45T/G in Adiponectin Gene and Restenosis after PCI

Making any conclusions on the grounds of single common SNP is generally not advisable in complex diseases, including IHD and its complications. On the other hand, the analyses based on a larger number of loci/polymorphisms are sometimes difficult to interpret and in the clinical setting, such analyses often lose their potential is patient screening and risk stratification, as it might be very difficult to make any clear, irreversible and justifiable decisions based purely on multiple loci analysis.

On the other hand, the analyses based exclusively on phenotypic data might be sometimes misleading as well, as variations in plasma levels of investigated proteins might occur both due to genetic constitution of the individual (e.g. might be constitutively increased) or due to external factors influencing disease progression and this might be vital information in the therapeutic decision-making.

Therefore, it remains to be established whether the increased adiponectin plasma levels observed in restenotic patients are due to specific individual genetic background (levels elevated/reduced already before the cardiac event) confounding with predisposition of the patient for the primary disease as well as for the worse PCI outcome or whether they are post-hoc elevated/decreased plasma levels as a pathophysiological consequence of the primary disease. After this is elucidated, adiponectin might become a useful diagnostic/prognostic marker for IHD.

None of the mentioned studies except a small study by our group [56] investigated the possible relationships between ADIPOQ polymorphism and restenosis after PCI. In the study, a total of 88 study subject undergoing repeating coronarography due to ischemic heart disease (IHD) were genotyped for +45T/G polymorphism in ADIPOQ gene. Although the study was apparently underpowered, we identified association of T allele with minimal lumen diameter, whereas the effect remained significant after all proper adjustments. The +45T/G polymorphism in ADIPOQ locus is not functional, it is a synonymous mutation that is highly likely to be in linkage disequilibrium with some other functional genetic loci responsible for an alteration in production of adiponectin or for the ability of adiponectin to polymerize, which affects its bioavailability and mode of action [57]. Still, this SNP was consistently reported to be associated with variety of pathological conditions, including CAD. The study undertaken in a cohort of 216 Afro-Caribbean T2D individuals revealed that +45T/G and hypoadiponectinemia were associated with CAD in the T2D subjects of predominantly African background and the authors report this effect to be isoform-specific [58] and it further confirms the outlined relationship between this polymorphisms and CAD and its associated complications, such as restenosis after PCI.

However, to make any relevant conclusions on prediction role of +45T/G on the risk of restenosis development after PCI, large population studies with appropriate study design and sufficient statistical power are crucial.

Conclusion

Undoubtedly, the studies presented so far have failed to strictly define precise physiologal and pathophysiological effects of adiponectin, either those related to adiponectin plasma/serum levels or particular genetic markers in ADIPOQ locus.

Although the number of studies investigating the potential of adiponectin plasma levels in restenosis prediction after PCI is rising, a very limited number of studies have attempted to investigate possible associations of variability in adiponectin gene with restenosis after PCI. However, this might be of great importance as some of the SNPs in ADIPOQ locus recently proved functional.

Associations between adiponectin plasma levels and presence and severity of CAD seem to be at least partially gender-specific, expressing increased strength in men compared to women. As the virtually only studie focused directly on restenosis risk in relation to adiponectin genetic marker (+45T/G) was apparently underpowered, replication of the data on larger population cohorts, different ethnic groups and using appropriate design and evaluation methodology is highly advisable.

Should these studies have been completed and the pathophysiological role of adiponectin in restenosis properly unravelled, adiponectin could play an important role as a marker of restenosis after PCI as well as a possible therapeutic target.

References

[1] Zhang Y, Proenca R, Maffei M, Barone M, Leopold L, Friedman JM. Positional cloning of the mouse obese gene and its human homologue. *Nature*. 1994;372:425–32.

[2] Scherer PE, Williams S, Fogliano M, Baldini G, Lodish HF. A novel serum protein similar to C1q, produced exclusively in adipocytes. *J. Biol. Chem.* 1995;270: 26746–26749.

[3] Nakano Y, Tobe T, Choi-Miura NH, Mazda T, Tomita M. Isolation and characterization of GBP28, a novel gelatin-binding protein purified from human plasma. *J. Biochem*. (Tokyo) 1996;120: 803–812.

[4] Maeda K, Okubo K, Shimomura I, Funahashi T, Matsuzawa Y, Matsubara K. cDNA cloning and expression of a novel adipose specific collagen-like factor, apM1 (adipose most abundant gene transcript 1) *Biochem. Biophys Res. Commun.* 1996;221: 286–289.

[5] Hu E, Liang P, Spiegelman BM. AdipoQ is a novel adipose-specific gene dysregulated in obesity. *J. Biol. Chem.* 1996;271:10697–10703.

[6] Ouchi N, Kihara S, Arita Y, Maeda K, Kuriyama H, Okamoto Y, Hotta K, Nishida M, Takahashi M, Nakamura T, Yamashita S, Funahashi T, Matsuzawa Y.. Novel modulator for endothelial adhesion molecules: adipocyte-derived plasma protein adiponectin. *Circulation* 1999;100:2473–6.

[7] Takahashi M, Arita Y, Yamagata K, Matsukawa Y, Okutomi K, Horie M, Shimomura I, Hotta K, Kuriyama H, Kihara S, Nakamura T, Yamashita S, Funahashi T, Matsuzawa Y. Genomic structure and mutations in adipose-specific gene, adiponectin. *Int. J. Obes Relat. Metab. Disord.* 2000;24: 861–868.

[8] Kobayashi H, Ouchi N, Kihara S, Walsh K, Kumada M, Abe Y, Funahashi T, Matsuzawa Y. Selective suppression of endothelial cell apoptosis by the high molecular weight form of adiponectin. *Circ. Res.* 2004 Mar 5;94(4):e27-31.

[9] Arita Y, Kihara S, Ouchi N, Maeda K, Kuriyama H, Okamoto Y, Kumada M, Hotta K, Nishida M, Takahashi M, Nakamura T, Shimomura I, Muraguchi M, Ohmoto Y, Funahashi T, Matsuzawa Y. Adipocyte-derived plasma protein adiponectin acts as a platelet-derived growth factor-BB-binding protein and regulates growth factor-induced common postreceptor signal in vascular smooth muscle cell. *Circulation* 2002;105: 2893–2898.

[10] Lam KS, Xu A. Adiponectin: protection of the endothelium. *Curr. Diabetes Rep.* 2005;5, 254–259.

[11] Trujillo ME, Scherer PE. Adiponectin: journey from an adipocyte secretory protein to biomarker of the metabolic syndrome. *J. Intern. Med.* 2005;257:167–175.

[12] Wang Y, Lam KS, Xu A. Adiponectin as a negative regulator in obesity-related mammary carcinogenesis. *Cell Res.*2007;17:280–282.

[13] Yamauchi T, Kamon J, Ito Y, Tsuchida A, Yokomizp T, Kita S, Sugiyama T, Miyagishi M, Hara K, Tsunoda M, Murakami K, Ohteki T, Uchida S, Takekawa S, Waki H, Tsuno NH, Shibata Y, Terauchi Y, Froguel P, Tobe K, Koyasu S, Taira K, Kitamura T, Shimizu T, Nagai R, Kadowaki T. Cloning of adiponectin receptors that mediate antidiabetic metabolic effects. *Nature* 2003;423: 762–769.

[14] Motoshima H, Wu X, Mahadev K, Goldstein BJ. Adiponectin suppresses proliferation and superoxide generation and enhances eNOS activity in endothelial cells treated with oxidized LDL. *Biochem. Biophys. Res. Commun.* 2004;315: 264–271.

[15] Pineiro R, Iglesias MJ, Gallego R, Raghay K, Eiras S, Rubio J, Dieguez C, Gualillo O, Gonzalez-Juanatey JR, Lago F. Adiponectin is synthesized and secreted by human and murine cardiomyocytes. *FEBS Lett.* 2005;579: 5163–5169.

[16] Yoon MJ, Lee GY, Chung JJ, Ahn YH, Hong SH, Kim JB. Adiponectin increases fatty acid oxidation in skeletal muscle cells by sequential activation of AMP-activated protein kinase, p38 mitogen-activated protein kinase, and peroxisome proliferator-activated receptor alpha. Diabetes. 2006;55(9):2562-70.

[17] Matsuda M, Shimomura I, Sata M, Arita Y, Nishida M, Maeda N, Kumada M, Okamoto Y, Nagaretani H, Nishizawa H, Kishida K, Komuro R, Ouchi N, Kihara S, Nagai R, Funahashi T, Matsuzawa Y. Role of adiponectin in preventing vascular stenosis. The missing link of adipo-vascular axis. *J. Biol. Chem.* 2002;4;277(40):37487-91.

[18] Ouedraogo R, Gong Y, Berzins B, Wu X, Mahadev K, Hough K, Chan L, Goldstein BJ, Scalia R. Adiponectin deficiency increases leukocyte-endothelium interactions via upregulation of endothelial cell adhesion molecules in vivo. *J. Clin. Invest.* 2007;117:1718–26.

[19] Okamoto Y, Kihara S, Ouchi N, Nishida M, Arita Y, Kumada M, Ohashi K, Sakai N, Shimomura I, Kobayashi H, Terasaka N, Inaba T, Funahashi T, Matsuzawa Y. Adiponectin reduces atherosclerosis in apolipoprotein E-deficient mice. *Circulation* 2002;106:2767–70.

[20] Yamauchi T, Kamon J, Waki H, Terauchi Y, Kubota N, Hara K, Mori Y, Ide T, Murakami K, Tsuboyama-Kasaoka N, Ezaki O, Akanuma Y, Gavrilova O, Vinson C, Reitman ML, Kagechika H, Shudo K, Yoda M, Nakano Y, Tobe K, Nagai R, Kimura S, Tomita M, Froguel P, Kadowaki T. The fat-derived hormone adiponectin reverses insulin resistance associated with both lipoatrophy and obesity. *Nat. Med.* 2001;7: 941–946.

[21] Ouchi N, Ohishi M, Kihara S, Funahashi T, Nakamura T, Nagaretani H, Kumada M, Ohashi K, Okamoto Y, Nishizawa H, Kishida K, Maeda N, Nagasawa A, Kobayashi H, Hiraoka H, Komai N, Kaibe M, Rakugi H, Ogihara T, Matsuzawa Y. Association of hypoadiponectinemia with impaired vasoreactivity. *Hypertension*. 2003;42(3):231-4.

[22] Ouchi N, Kihara S, Arita Y, Maeda K, Kuriyama H, Okamoto Y, Hotta K, Nishida M, Takahashi M, Nakamura T, Yamashita S, Funahashi T, Matsuzawa Y. Novel modulator for endothelial adhesion molecules: adipocyte-derived plasma protein adiponectin. *Circulation* 100: 2473–2476, 1999.

[23] Maeda N, Takahashi M, Funahashi T, Kihara S, Kishizawa H, Kishida K, Nagaretani H, Matsuda M, Komuro R, Ouchi N, Kuriyama H, Hotta K, Nakamura T, Shimomura I, Matsuzawa Y. PPAR ligands increase expression and plasma concentrations of adiponectin, an adipose-derived protein. *Diabetes* 2001;50: 2094–2099.

[24] Kumada M, Kihara S, Ouchi N, Kobayashi H, Okamoto Y, Ohashi K, Maeda K, Nagaretani H, Kishida K, Maeda N, Nagasawa A, Funahashi T, Matsuzawa Y. Adiponectin specifically increased tissue inhibitor of metalloproteinase-1 through interleukin-10 expression in human macrophages. *Circulation* 109: 2046–2049, 2004.

[25] Wolf AM, Wolf D, Rumpold H, Enrich B, Tilg H.Adiponectin induces the anti inflammatory cytokines IL-10 and IL-1RA in human leukocytes. Biochem Biophys Res Commun. 2004 Oct 15;323(2):630-5.

[26] Fasshauer M, Kralisch S, Klier M, Lossner U, Bluher M, Klein J, Paschke R. Adiponectin gene expression and secretion is inhibited by interleukin-6 in 3T3-L1 adipocytes. *Biochem. Biophys. Res. Commun.* 2003;301:1045–1050.

[27] Kato H, Kashiwagi H, Shiraga M, Tadokoro S, Kamae T, Ujiie H, Honda S, Miyata S, Ijiri Y, Yamamoto J, Maeda N, Funahashi T, Kurata Y, Shimomura I, Tomiyama Y, Kanakura Y.Adiponectin acts as an endogenous antithrombotic factor. *Arterioscler. Thromb. Vasc. Biol* 2006;26:224–230.

[28] Hoo RL, Chow WS, Yau MH, Xu A, Tso AW, Tse HF, Fong CH, Tam S, Chan L, Lam KS. Adiponectin mediates the suppressive effect of rosiglitazone on plasminogen activator inhibitor-1 production.*Arterioscler Thromb. Vasc. Biol.* 2007;27(12):2777-82.

[29] Chen H, Montagnani M, Funahashi T, Shimomura I, Quon MJ. Adiponectin stimulates production of nitric oxide in vascular endothelial cells. *J. Biol. Chem* 2003;278: 45021–45026.

[30] Hattori Y, Suzuki M, Hattori S, Kasai K. Globular adiponectin upregulates nitric oxide production in vascular endothelial cells. *Diabetologia* 2003;46:1543–1549.

[31] Ouedraogo R, Wu X, Xu SQ, Fuchsel L, Motoshima H, Mahadev K, Hough K, Scalia R, Goldstein BJ. Adiponectin suppression of high-glucose-induced reactive oxygen species in vascular endothelial cells: evidence for involvement of a cAMP signaling pathway. *Diabetes* 2006;55:1840–6.

[32] Tao L, Gao E, Jiao X, Yuan Y, Li S, Christopher TA, Lopez BL, Koch W, Chan L, Goldstein BJ, Ma XL. Adiponectin cardioprotection after myocardial ischemia/reperfusion involves the reduction of oxidative/nitrative stress. *Circulation* 2007;115:1408–1416.

[33] Kase H, Hattori Y, Jojima T, Okayasu T, Tomizawa A, Suzuki K, Banba N, Monden T, Satoh H, Akimoto K, Kasai K.Globular adiponectin induces adhesion molecule expression through the sphingosine kinase pathway in vascular endothelial cells. *Life Sci.* 2007;81:939–943.

[34] Ouchi N, Kihara S, Arita Y, Nishida M, Matsuyama A, Okamoto Y, Ishigami M, Kuriyama H, Kishida K, Nishizawa H, Hotta K, Muraguchi M, Ohmoto Y, Yamashita S, Funahashi T, Matsuzawa Y. Adipocyte-derived plasma protein, adiponectin, suppresses lipid accumulation and class A scavenger receptor expression in human monocytederived macrophages. *Circulation* 2001;103: 1057–1063.

[35] Yokota T, Oritani K, Takahashi I, Ishikawa J, Matsuyama A, Ouchi N, Kihara S, Funahashi T, Tenner AJ, Tomiyama Y, Matsuzawa Y. Adiponectin, a new member of the family of soluble defense collagens, negatively regulates the growth of myelomonocytic progenitors and the functions of macrophages. *Blood.* 2000;96:1723–1732.

[36] Iglseder B, Mackevics V, Stadlmayer A, Tasch G, Ladurner G, Paulweber B. Plasma adiponectin levels and sonographic phenotypes of subclinical carotid artery atherosclerosis: data from the SAPHIR Study. *Stroke.* 2005;36:2577-82.

[37] Dullaart RP, Kappelle PJ, Dallinga-Thie GM.Carotid intima media thickness is associated with plasma adiponectin but not with the leptin:adiponectin ratio independently of metabolic syndrome. *Atherosclerosis*. 2010 Mar 25.

[38] Shimetani N.Evaluation of atherosclerosis using data on serum lipids, serum adiponectin, carotid ultrasonography, and pulse wave velocity: implications for preventive medical care. *Rinsho Byori.* 2008 Oct;56(10):877-86.

[39] Iwashima Y, Horio T, Suzuki Y, Kihara S, Rakugi H, Kangawa K, Funahashi T, Ogihara T, Kawano Y. Adiponectin and inflammatory markers in peripheral arterial occlusive disease. *Atherosclerosis.* 2006 Oct;188(2):384-90

[40] Golledge J, Leicht AS, Crowther RG, Glanville S, Clancy P, Sangla KS, Spinks WL, Quigley F. Determinants of endothelial function in a cohort of patients with peripheral artery disease.*Cardiology.* 2008;111(1):51-6.

[41] Kiris I, Tekin I, Yesildag A, Vural H, Oyar O, Sirin B, Okutan H, Ibrisim E. Inverse relationship between adiponectin levels and subclinical carotid atherosclerosis in patients undergoing coronary artery bypass grafting. Int Heart J. 2006 Nov;47(6):855-66

[42] Eyileten Z, Yilmaz MI, Kaya K, Akar AR, Kahraman D, Bingol S, Uysalel A, Ozyurda U. Coronary artery bypass grafting ameliorates the decreased plasma adiponectin level in atherosclerotic patients. *Tohoku J. Exp. Med.* 2007 Sep;213(1):71-7.

[43] Shioji K, Moriwaki S, Takeuchi Y, Uegaito T, Mutsuo S, Matsuda M. Relationship of serum adiponectin level to adverse cardiovascular events in patients who undergo percutaneous coronary intervention. *Circ. J.* 2007 May;71(5):675-80.
[44] Inami N, Nomura S, Shimazu T, Manabe K, Kimura Y, Iwasaka T. Adiponectin incompletely prevent MCP-1-dependent restenosis after percutaneous coronary intervention [corrected] in patients with coronary artery disease. *J. Thromb. Thrombolysis.* 2007 Dec;24(3):267-73.
[45] Kitta Y, Takano H, Nakamura T, Kodama Y, Umetani K, Fujioka D, Saito Y, Kawabata K, Obata JE, Mende A, Kobayashi T, Kugiyama K. Low adiponectin levels predict late in-stent restenosis after bare metal stenting in native coronary arteries. *Int. J. Cardiol.* 2008;17;131(1):78-82.
[46] Ambrosi P, Oddoze C, Nicolay A, Penet-Lorec AM, Riberi A, Métras D, Portugal H, Habib G. Plasma adiponectin in heart transplant recipients. *Clin. Transplant.* 2009 Jan;23(1):83-8.
[47] Pischon T, Girman CJ, Hotamisligil GS, Rifai N, Hu FB, Rimm EB. Plasma adiponectin levels and risk of myocardial infarction in men. *JAMA*. 2004;14;291 (14): 1730-7.
[48] Sattar N, Wannamethee G, Sarwar N, Tchernova J, Cherry L, Wallace AM, Danesh J, Whincup PH. Adiponectin and coronary heart disease: a prospective study and meta-analysis. *Circulation* 2006;114:623–9.
[49] Lawlor DA, Davey Smith G, Ebrahim S, Thompson C, Sattar N.Plasma adiponectin levels are associated with insulin resistance, but do not predict future risk of coronary heart disease in women. *J. Clin. Endocrinol. Metab.* 2005;90:5677-5683.
[50] Iglseder B, Mackevics V, Stadlmayer A, Tasch G, Ladurner G, Paulweber B. Plasma adiponectin levels and sonographic phenotypes of subclinical carotid artery atehrosclerosis. *Stroke* 2005;36: 2577–2582.
[51] Kojima S, Funahashi T, Maruyoshi H, Honda O, Sugiyama S, Kawano H, Soejima H, Miyamoto S, Hokamaki J, Sakamoto Y, Yoshimura M, Kitagawa A, Matsuzawa Y, Ogawa H. Levels of the adipocyte-derived plasma protein, adiponectin, have a close relationship with atheroma. *Thromb. Res.* 2005;115: 483–490.
[52] Corrado E, Rizzo M, Coppola G, Fattouch K, Novo G, Marturana I, Ferrara F, Novo S. An update on the role of markers of inflammation in atherosclerosis. *J. Atheroscler Thromb.* 2010;17(1):1-11.
[53] Matsubara M, Namioka K, Katayose S. Decreased plasma adiponectin concentrations in women with low grade C-reactive protein elevation. *Eur. J. Endocrinol.* 2003;148: 657–662.
[54] Pischon T, Pai JK, Manson JE, Hu FB, Rexrode KM, Hunter D, Rimm EB. Single nucleotide polymorphisms at the adiponectin locus and risk of coronary heart disease in men and women. *Obesity* (Silver Spring). 2007;15(8):2051-60.
[55] Vasseur F, Caeyseele T, Barat-Houari M, Lobbens S, Meirhaeghe A, Meyre D, Froguel P, Amouyel P, Helbecque N. Concordance of two multiple analytical approaches demonstrate that interaction between BMI and ADIPOQ haplotypes is a determinant of LDL cholesterol in a general French population. *J. Hum. Genet.* 2010, e-pub Feb 26.
[56] Bienertova-Vasku J, Bienert P, Hlinomaz O, Vasku A. Common polymorphism +45T/G in adiponectin gene as potential modulator of in-stent restenosis development. *Int. J. Cardiol.* 2009, e-pub Dec 30.

[57] Comuzzie AG, Funahashi T, Sonnenberg G, Martin LJ, Jacob HJ, Black AE, Maas D, Takahashi M, Kihara S, Tanaka S, Matsuzawa Y, Blangero J, Cohen D, Kissebah A. The genetic basis of plasma variation in adiponectin, a global endophenotype for obesity and the metabolic syndrome. *J. Clin. Endocrinol. Metab.* 2001;86:4321–5.

[58] Foucan L, Ezourhi N, Maimaitiming S, Hedreville S, Inamo J, Atallah A, Bangou-Bredent J, Aubert R, Chout R, Fumeron F, Donnet JP, Marre M. Adiponectin Multimers and ADIPOQ T45G in Coronary Artery Disease in Caribbean Type 2 Diabetic Subjects of African Descent. *Obesity* (Silver Spring). 2009 Dec 3.

In: Adiponectin: Production, Regulation and Roles in Disease ISBN: 978-1-61942-165-3
Editor: Masayoshi Yamaguchi

Chapter IV

ADIPONECTIN AND MACROPHAGE

***Sumio Akifusa*[*,1] *and Yoshihisa Yamashita*[2]**
[1]Department of Health Management, School of Oral Health Sciences,
Kyushu Dental Coellege, Fukuoka, Japan
[2]Section of Preventive and Public Health Dentistry,
Division of Oral Health, Growth, and Development,
Faculty of Dental Science, Kyushu University, Fukuoka, Japan

ABSTRACT

Adipose tissue secretes various bioactive soluble mediator called adipokines. In excess adipose tissue, increase of total amount of proinflammatory adipokines, such as TNF-α and IL-6 are observed, indicating that obesity is associated with a state of low-grade, chronic inflammation. Adiponectin, however, mature adipocyte-derived peptide, shows negative correlation with body mass index and decreases its plasma level in obese. As clinical feature, hypoadiponectinaemia is closely related with onset of metabolic syndrome, cardiovascular disease, and type II diabetes. Administration of adiponectin improves insulin resistance and/or inflammatory condition in vivo. Thus, adiponectin possessing anti-inflammatory and anti-metabolic dysfunctional characterization is expected to be used for treatment and/or preventative agent. In this review, we summarized the role of adiponectin on metabolic and immune function, in particular function of macrophage.

Keywords: adiponectin, macrophage

INTRODUCTION

Obesity is a severe health problem in developed countries. According to the U.S. Centers for Disease Control and Prevention, high-calorie diet, combined with physical inactivity, is the second largest cause of death, with 400,000 people dying every year from obesity-related

* E-mail address: r11akifusa@fa.dent.kyu-dent.ac.jp

causes. This figure is second only to the annual number of deaths from cigarette smoking (435,000) [1]. Adipose tissue that is enlarged because of fat accumulation secretes a wide variety of bioactive substances. Thus, adipose tissue is important not only for energy storage but is also as a large endocrine organ. Among the bioactive substances secreted by adipocytes are a group of peptides, collectively termed “adipokines,” that possess cytokine-like activities. It has been shown that these peptides are strongly associated with the symptoms of obesity, such as the metabolic syndrome. In recent years, because of the increased concentration of adipokines present in obesity, particularly of the visceral variety, obesity has been recognized as a state of weak, chronic inflammation. The term “metaflammation,” meaning inflammation induced by metabolic disorders, has also been used to characterize obesity [2]. To date, a large number of adipokines have been identified. Inflammatory cytokines, such as tumor necrosis factor-α (TNF-α) and interleukin-6 (IL-6), which are also secreted from adipose tissue, affect energy metabolism by inhibiting the action of insulin and the metabolism of glucose. In addition, recent research has identified angiotensinogen, apelin, chemerin, leptin, obestatin, omentin, PAI-1, resistin, vaspin, visfatin, and others as adipokines. Researchers are now actively studying the biological activity of these adipokines and the relationship between the concentration of adipokines and diseases [3,4]. There are some adipokines of which the expression is decreased when fat is stored. The adipokine adiponectin level is inversely correlated with body mass index (BMI) and has been implicated in metabolism and immune system function. Because adiponectin is also inversely correlated with insulin resistance, the media often portray adiponectin as a “longevity hormone.” Several researches have been conducted on the potential of adiponectin replacement therapy. There have also been researches on the effects of thiazolidinediones [5] and glimepiride [6], which are known to induce adiponectin expression. In the present report, we review the biological activities of adiponectin and introduce some recent reports regarding the relationship between adiponectin and the immune system, with a focus on macrophage function.

ADIPONECTINE AND ITS RECEPTOR

Adiponectin is a 30-kDa hormone-like peptide that contains a globular domain at the C-terminus and a collagen-like domain at the N-terminus, similar to complement C1q. It is secreted from adipocytes, and its concentration in the bloodstream ranges from 5 to 10 μg/ml. This concentration is much higher than those of other cytokines. Adiponectin forms the following isoforms: (1) a low-molecular-weight trimer, (2) a middle-molecular-weight hexamer, (3) high-molecular-weight 12- to 18-mers, and (4) low-molecular-weight isoforms that bind to albumin and albumin isoforms [7]. It has been suggested that every adiponectin isoform exhibits a distinct type of activity [8]. Data obtained using an ELISA assay that only detects high-molecular-weight isoforms indicate that the high-molecular-weight isoforms are more closely related to insulin resistance, the metabolic syndrome, and circulation diseases than are the other isoforms [9]. The polymerization of adiponectin occurs in the Golgi apparatus, along with posttranslational modifications such as glycosylation and proteolysis of the signal peptide [10]. A globular type of adiponectin (gAd) that has a truncated collagen-like domain in the low-molecular-weight isoform has also been reported. Because both the full-length and globular isoforms of adiponectin exhibit similar activities in macrophages and monocytes (although some reports have shown that the globular isoform has more potent activity), the difference between these two isoforms is still unclear. However, a report that lymphoid-type esterase is involved in the cleavage of the N-terminal collagen-like domain suggests that globular adiponectin is related to inflammation [11].

In obesity, the pressure from enlarged adipose tissue causes the peripheral environment to become hypoxic. In this state, the expression of adiponectin is inhibited by hypoxia-induced factor α (HIF-α) [12]. In addition, the expression of adiponectin is inhibited by TNF-α, IL-6, and IL-18, cytokines that are increased in obesity and type 2 diabetes. These phenomena could explain the clinical observation that serum adiponectin concentrations are decreased in obese patients. Adiponectin is involved in energy metabolism by inhibiting gluconeogenesis in the liver, promoting fatty acid oxidation in the liver and skeletal muscle, increasing glucose uptake in skeletal muscle, and enhancing insulin secretion.

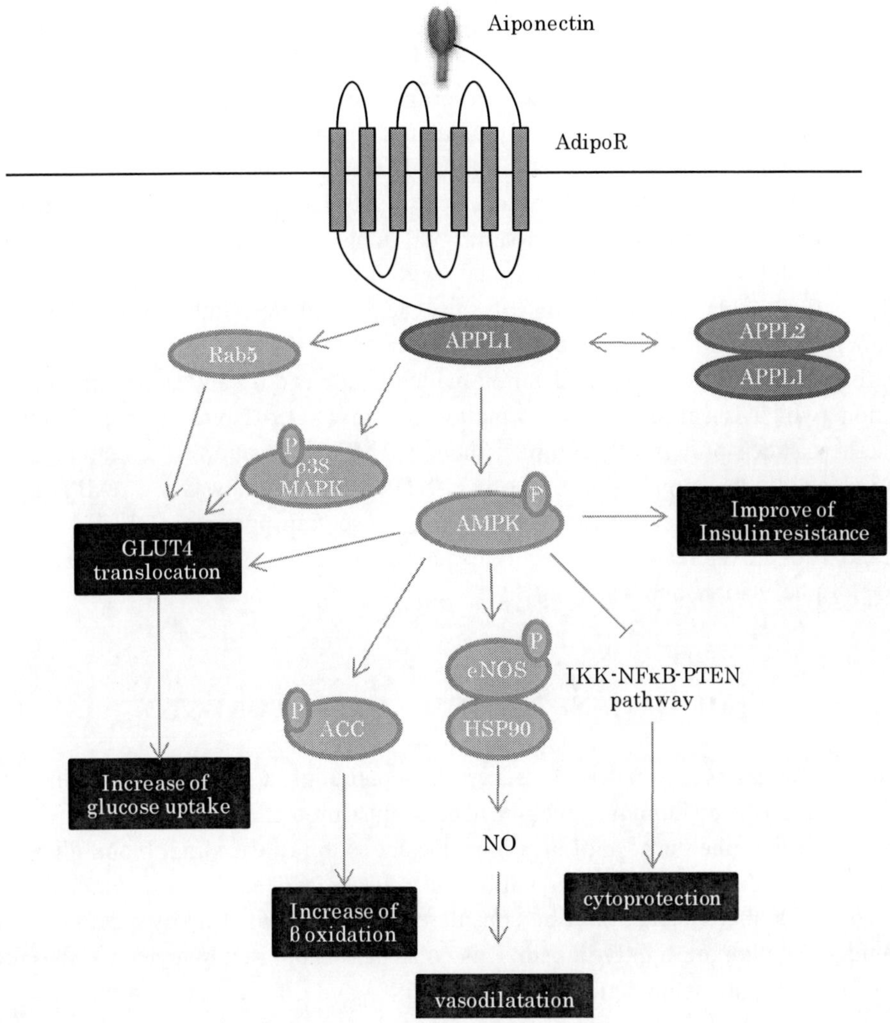

Figure 1. The mechanism of adiponectin action. Adiponectin performs a wide variety of functions by binding to its receptors (AdipoR1 and AdipoR2) and to cytosolic APPL1. Adiponectin promotes glucose uptake, fatty acid oxidation, and vasodilatation. It also protects cells from cytokines and restores decreased insulin sensitivity (reference 20).

It is well known that caloric restriction is associated with longevity. The concentration of serum adiponectin is higher in people older than 95 years and their descendants [13]. In addition, polymorphisms in the adiponectin gene are associated with male longevity [14]. Recently, adiponectin-inducing signals closely related with activation of AMPK and/or Sir1, key molecules for calorie restriction longevity [15] [16] [17]. Taken together, these studies suggest that the effects of calorie consumption on adiponectin are related to longevity.

Kadowaki's group isolated the adiponectin receptors, AdipoR1 and AdipoR2, using cell sorter system [18]. Although AdiopR1 is widely expressed, AdipoR2 is enriched only in the liver. The adiponectin receptor was expected to be a G-protein-coupled receptor (GPCR) with seven transmembrane domains, but the G-protein coupling activity of AdipoR1 and AdipoR2 is still not clear. In addition, because their N-termini are localized to the cytosol and their C-termini are outside the cell, their topology is opposite to that of a classical GPCR. Whereas AdipoR1 has a high affinity for both the full-length and globular forms of adiponectin, AdipoR2 has a low affinity for globular adiponectin. APPL (adaptor protein containing PH domain, PTB domain, and leucine zipper motif) has been identified as an intracellular biding partner of AdipoR [19]. APPL has two isoforms, APPL1 and APPL2 [20], which form a heterodimer and positively and negatively regulate adiponectin signaling, respectively. In the steady state, APPL2 binds to the N-terminus of AdipoR. Upon adiponectin binding to AdipoR, the APPL heterodimer dissociates, and APPL1 is recruited to the intracellular domain of AdipoR1, leading to downstream signal transduction [19].

As shown in Figure 1, a variety of signals with diverse outcomes are transmitted through APPL1, including (1) the activation of fatty acid oxidation induced by the activation (phosphorylation) of AMP-activated protein kinase (AMPK) and the inactivation of acetyl-CoA carboxylase (ACC); (2) vasodilatation, which is induced by increased nitric oxide (NO) production (which is caused by the formation of the eNOS-Hsp90 [epithelial nitric oxide synthase-heat shock protein 90] complex though AMPK activation); (3) cellular protective effects induced by the suppression of the IKK-NF-κB-PTEN pathway via AMPK activation; and (4) the localization of insulin-responsive glucose transporter 4 (GLUT4) to the cell membrane and the increased glucose transport induced by AMPK phosphorylation, p38 MAPK phosphorylation, and Rab5 [20].

THE IMMUNE SYSTEM AND ADIPONECTIN

Human homeostasis is based on starvation conditions. Considering insulin resistance from the perspective of immune function, which must protect the body from infection even during malnutrition, the purpose of insulin resistance is to preserve high blood glucose levels to maintain immune function by inhibiting muscle glucose uptake under starvation conditions. Thus, this system could be considered a biophylaxis. Energy metabolism/storage and immune function are tightly linked. This concept is supported by evidence that the liver, adipose tissue, and hematopoietic system are derived and differentiated from a tissue that is evolutionarily similar to the insect fat body [2]. For example, adipose tissue-derived adiponectin plays an important role in liver recovery [21]. Macrophage invasion has been observed in enlarged adipose tissue [22]. It has been suggested that these macrophages are derived not only from bone marrow but also from pre-adipocytes or mesenchymal stem cells

[23]. In addition, adiponectin functions as a growth factor for hematopoietic stem cells. These observations indicate the evolutionary linkage between adipose tissue and the hematopoietic system [24].

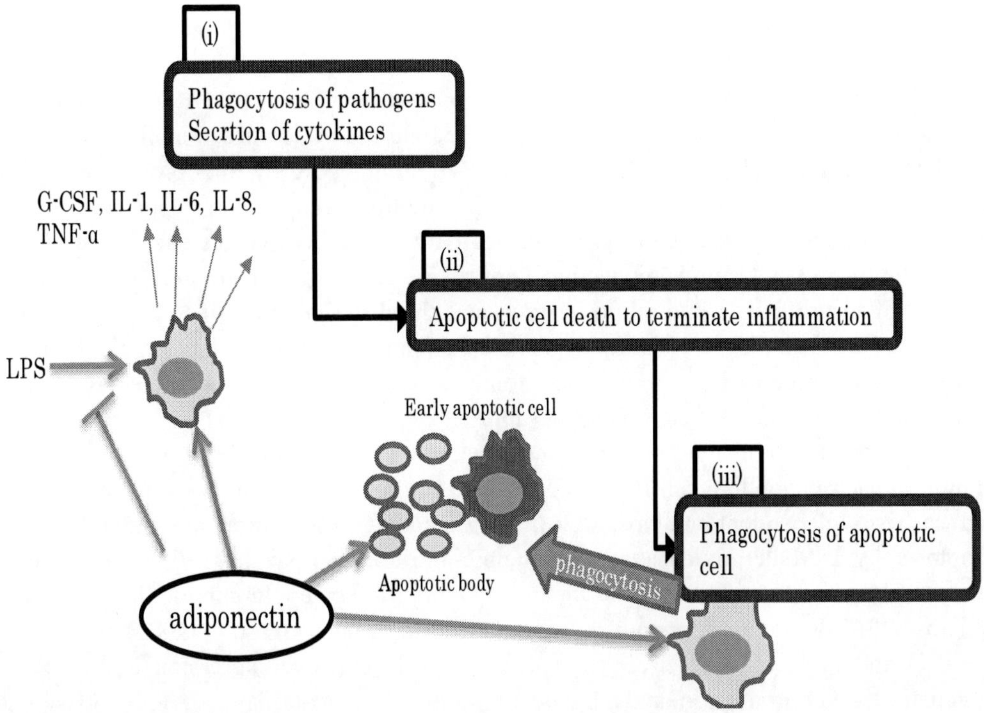

Figure 2. The relationship between adiponectin and macrophage function. Adiponectin affects the following macrophage functions: (i) the phagocytosis of pathogens and the secretion of cytokines, (ii) the induction of apoptosis, and (iii) the phagocytosis of apoptotic cells.

Changes in serum concentrations of adiponectin have also been implicated in the pathogenesis of atherosclerosis [25]. Decreased serum adiponectin levels result in the increased expression of adhesion molecules in vascular endothelial cells, which facilitates monocyte attachment to the endothelial cells. These effects enhance plaque formation and cytokine secretion from macrophages that have invaded the vascular endothelial cells [26]. Adiponectin knockout (KO) mice exhibit metabolic abnormalities, such as decreased serum lipid clearance and insulin resistance accompanied by reduced IRS-1-associated PI3K activity. These mice also have immune defects, such as increased serum TNF-α concentration [27]. In addition, adiponectin represses lipopolysaccharide (LPS)-induced NF-κB signaling in monocytes and macrophages. Furthermore, adiponectin represses the activity of natural killer cells induced by IL-2 and interferon-γ (IFN-γ) [28]. Based on these reports, adiponectin is thought to be an anti-inflammatory adipokine. However, it has been reported that in autoimmune diseases (e.g., rheumatoid arthritis, systemic lupus erythematous, type 1 diabetes, inflammatory bowel disease), plasma level of adiponectin is locally and/or systemically increased [29]. This observation suggests that adiponectin may overactivate the immune response, suggesting the regulation of the adiponectin serum concentration is critical to keep a sound immune system.

Macrophage and Adiponectin

Macrophages express both AdipoR1 and AdipoR2, and adiponectin has been shown to exhibit a wide variety of activities against macrophages (Figure 2). Adiponectin inhibits signaling from the LPS receptor toll-like receptor 4 (TLR4) by suppressing NF-κB and extracellular signal-regulated kinase 1/2 (ERK1/2) activity. This inhibition results in decrease of LPS-induced TNF-α secretion and NO production [30,31]. The counteraction of adiponectin against LPS stimulation is mediated through the direct inhibition of TLR4 signaling and the indirect inhibition of TNF-α stimulation by the promotion of anti-inflammatory cytokine IL-10 secretion [32]. In addition, adiponectin suppresses the LPS-induced production of reactive oxygen species (ROS) in macrophages such as the Kupffer cells in the liver [33]. Using adiponectin KO mice, it was shown that adiponectin inhibited the invasion of macrophages into adipose tissue and the differentiation of M1 macrophages in adipose tissue [34]. These reports suggest that adiponectin represses macrophage activity. However, other studies have shown that adiponectin alone could induce the production of inflammatory cytokines in macrophages (Table 1) [35].

Adiponectin inhibits macrophage growth by inducing apoptosis in pre-macrophages [31]. In mouse macrophage like cell line RAW264 cells, adiponectin treatment induces caspase-3- and caspase-9-dependent apoptosis [36]. ROS and NO production are induced prior to apoptosis by NADPH oxidases and inducible nitric oxide synthase (iNOS) production, respectively; therefore, blocking ROS and NO production could inhibit the induction of apoptosis [36]. In addition, apoptosis is dependent on protein kinase C (PKC) activity. ROS and NO are inhibited by Bcl-2 overexpression [37]. The overexpression of SOCS3 also inhibits ROS, NO production, and adiponectin-induced apoptosis. Thus, JAK-STAT signaling is involved in adiponectin-induced apoptosis (Figure 3) [38].

Table 1. Genes induced by adiponetin stimulation

genes	Expression ratio	
	30 min	60 min
Colony-stimulating factor 3 (granulocyte)	7.0	104.0
Interleukin 6	1.0	48.5
Colony-stimulating factor 2 (granulocyte–macrophage)	6.5	26.0
Interleukin 1 alpha	0.9	22.6
Phospholipase C, zeta 1	4.3	21.1
C1q and tumor necrosis factor-related protein 3	1.7	19.7
Chemokine (C–X–C motif) ligand 2	4.9	17.1
Tumor necrosis factor alpha	5.3	8.0
Interleukin 10	1.2	2.3

Adiponectin activates a wide variety of cellular signaling pathways. Treatment with gAd induces mTOR phosphorylation and activation [39]. In addition, treatment with the mTOR inhibitor rapamycin inhibits gAd-induced ROS and NO production. Since mTOR activity is related to AMPK, treatment with the AMPK activators 2-deoxy-D-glucose and A-769662 inhibits gAd-induced mTOR phosphorylation. Diacylglycerol kinase (DGK), which inhibits PKC, also plays an important role in regulating NADPH oxidases activity. It has also been reported that treatment with a DGK inhibitor or DGKα RNAi inhibits gAd-induced ROS and NO production [40]. Adiponectin has been shown to regulate calcium influx in a calmodulin-dependent fashion in myoblasts [15]. In macrophages, adiponectin also affects intracellular calcium movement [41].

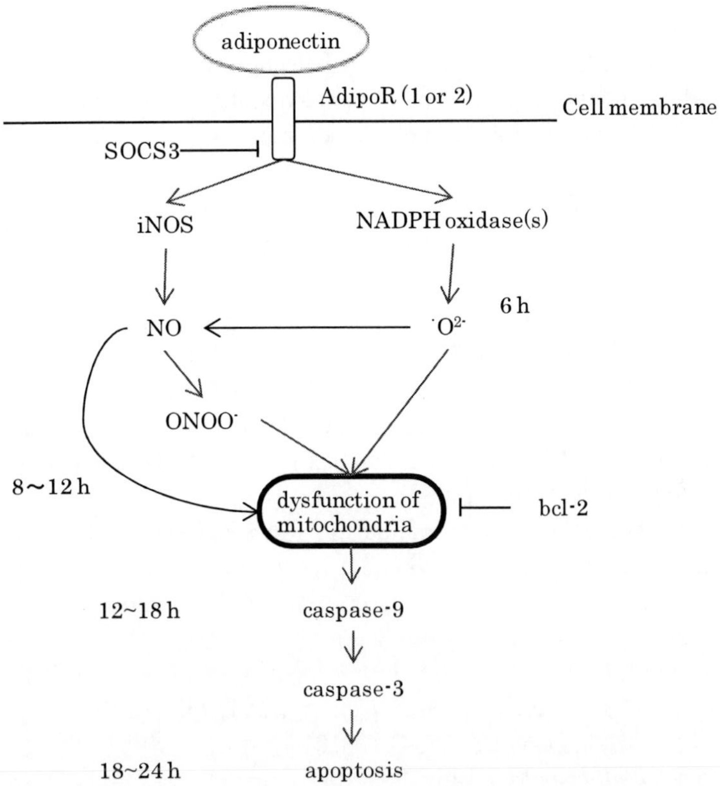

Figure 3. Intracellular signaling in apoptotic macrophages is induced by adiponectin. ROS and NO damage mitochondria, resulting in caspase-dependent apoptosis. This apoptotic process is inhibited by blocking iNOS-induced NO production and NADPH oxidases-induced ROS.

When intracellular calcium is depleted by treatment with BAPTA-AM, adiponectin-induced ROS and NO production are inhibited. In contrast, when the cells are treated with the NADPH oxidases inhibitor diphenyleneiodonium chloride, gAd-induced calcium oscillation is inhibited. These observations suggest that calcium movement is related to the presence of intracellular ROS. Furthermore, blocking calmodulin activity also inhibits gAd-induced NO production [41].

Macrophages remove apoptotic cells by phagocytosis. Adiponectin enhances phagocytosis by binding to calreticulin/CD91, which is expressed by the cell membranes of macrophages [42]. Because defects in this phagocytic process can result in atherosclerosis, it has been suggested that one of the anti-atherosclerotic roles of adiponectin is to remove cells in the early stages of apoptosis. Meanwhile, adiponectin also inhibits the LPS-induced phagocytosis of macrophages [31]. Furthermore, the repressive effects of adiponectin on LPS-induced cytokine secretion from macrophages raise the possibility that adiponectin inhibits the infection-induced overactivation of immune function.

Conclusion

Here, we reviewed the effects of adiponectin on immune function with a focus on the role of macrophages. Adiponectin level is not only a biomarker indicative of the metabolic syndrome, obesity-related diseases (such as diabetes), and control of the pathophysiology of metaflammation; it also has great potential for the treatment of metaflammation because of its repressive effects on inflammation. Thus, clarification of the roles of adiponectin in physiological and pathophysiological conditions may contribute to developing treatment and prevention methods for obesity and metabolic syndrome.

References

[1] A. H. Mokdad, J. S. Marks, D. F. Stroup and J. L. Gerberding, *JAMA* 291, 1238 (2004).

[2] G. S. Hotamisligil, *Nature* 444, 860 (2006).

[3] B. Antuna-Puente, B. Feve, S. Fellahi and J. P. Bastard, *Diabetes Metab.* 34, 2 (2008).

[4] K. Rabe, M. Lehrke, K. G. Parhofer and U. C. Broedl, *Mol. Med.* 14, 741 (2008).

[5] J. P. Whitehead, A. A. Richards, I. J. Hickman, G. A. Macdonald and J. B. Prins, *Diabetes Obes Metab.* 8, 264 (2006).

[6] Y. Kanda, M. Matsuda, K. Tawaramoto, F. Kawasaki, M. Hashiramoto, M. Matsuki and K. Kaku, *Diabetes Res. Clin. Pract* 81, 13 (2008).

[7] T. Kadowaki, T. Yamauchi, N. Kubota, K. Hara, K. Ueki and K. Tobe, *J. Clin. Invest.* 116, 1784 (2006).

[8] F. Simpson and J. P. Whitehead, *Int. J. Biochem. Cell Biol.* 42, 785 (2009).

[9] F. Ziemke and C. S. Mantzoros, *Am. J. Clin. Nutr.* 91, 258S (2010).

[10] M. Liu and F. Liu, *Biochem. J.* 425, 41 (2010).

[11] H. Waki et al., *Endocrinology* 146, 790 (2005).

[12] J. Ye, *Int. J. Obes (Lond)* 33, 54 (2009).

[13] G. Atzmon et al., *J. Gerontol A Biol. Sci. Med. Sci.* 63, 447 (2008).

[14] O. F. Khabour, F. S. Mesmar, M. A. Alatoum, M. Y. Gharaibeh and K. H. Alzoubi, *Aging Male* 13, 188

[15] M. Iwabu et al., *Nature* 464, 1313 (2010).

[16] Z. Shen, X. Liang, C. Q. Rogers, D. Rideout and M. You, *Am. J. Physiol. Gastrointest Liver Physiol.* 298, G364 (2010).

[17] M. P. Gillum, D. M. Erion and G. I. Shulman, *Trends Mol Med* (2010).

[18] T. Yamauchi et al., *Nature* 423, 762 (2003).

[19] . Wang et al., *J. Biol. Chem.* 284, 31608 (2009).

[20] S. S. Deepa and L. Q. Dong, *Am. J. Physiol. Endocrinol. Metab.* 296, E22 (2009).

[21] P. Mandal, P. H. Park, M. R. McMullen, B. T. Pratt and L. E. Nagy, *Hepatology* 51, 1420 (2010).

[22] S. P. Weisberg, D. McCann, M. Desai, M. Rosenbaum, R. L. Leibel and A. W. Ferrante, Jr., *J. Clin. Invest* 112, 1796 (2003).

[23] H. Tilg and A. R. Moschen, *Clin. Sci. (Lond)* 114, 275 (2008).

[24] L. DiMascio, C. Voermans, M. Uqoezwa, A. Duncan, D. Lu, J. Wu, U. Sankar and T. Reya, *J. Immunol.* 178, 3511 (2007).

[25] C. Lara-Castro, Y. Fu, B. H. Chung and W. T. Garvey, *Curr. Opin. Lipidol.* 18, 263 (2007).
[26] P. Kougias, H. Chai, P. H. Lin, Q. Yao, A. B. Lumsden and C. Chen, *J. Surg. Res.* 126, 121 (2005).
[27] N. Maeda et al., *Nat Med.* 8, 731 (2002).
[28] K. Y. Kim et al., *J. Immunol.* 176, 5958 (2006).
[29] G. Fantuzzi, *J. Allergy Clin. Immunol.* 121, 326 (2008).
[30] N. Yamaguchi, J. G. Argueta, Y. Masuhiro, M. Kagishita, K. Nonaka, T. Saito, S. Hanazawa and Y. Yamashita, *FEBS Lett* 579, 6821 (2005).
[31] T. Yokota et al., *Blood* 96, 1723 (2000).
[32] P. H. Park, H. Huang, M. R. McMullen, P. Mandal, L. Sun and L. E. Nagy, *J. Biol. Chem.* 283, 26850 (2008).
[33] P. H. Park, V. Thakur, M. T. Pritchard, M. R. McMullen and L. E. Nagy, *J. Gastroenterol .Hepatol.* 21 Suppl 3, S30 (2006).
[34] K. Ohashi et al., *J. Biol.Chem.* 285, 6153 (2010).
[35] N. Kamio, S. Akifusa, N. Yamaguchi and Y. Yamashita, *Mol. Cell Endocrinol.* 292, 20 (2008).
[36] S. Akifusa, N. Kamio, Y. Shimazaki, N. Yamaguchi and Y. Yamashita, *Free Radic. Biol. Med.* 45, 1326 (2008).
[37] S. Akifusa, N. Kamio, Y. Shimazaki, N. Yamaguchi, T. Nishihara and Y. Yamashita, *Free Radic Biol. Med.* 46, 1308 (2009).
[38] S. Akifusa, N. Kamio, Y. Shimazaki, N. Yamaguchi, K. Nonaka and Y. Yamashita, *J. Cell Biochem.* 111, 597 (2010).
[39] A. Fujimoto, S. Akifusa, N. Kamio, T. Hirofuji, K. Nonaka and Y. Yamashita, *Free Radic. Res.* 44, 128 (2010).
[40] N. Kamio, S. Akifusa and Y. Yamashita, *Free Radic. Res.* 45, 336 (2011).
[41] S. Akifusa, N. Kamio, Y. Shimazaki, N. Yamaguchi and Y. Yamashita, *Biochem. Biophys. Res. Commun.* 381, 649 (2009).
[42] Y. Takemura, N. Ouchi, R. Shibata, T. Aprahamian, M. T. Kirber, R. S. Summer, S. Kihara and K. Walsh, *J. Clin. Invest* 117, 375 (2007).

In: Adiponectin: Production, Regulation and Roles in Disease
Editor: Masayoshi Yamaguchi
ISBN: 978-1-61942-165-3

Chapter V

ADIPONECTIN AND CHRONIC LIVER DISEASES: AN UPDATE

Yu Wang[1,2,*], Karen S. L. Lam[2,3] and Aimin Xu[1,2,3]
[1]Department of Pharmacology and Pharmacy,
[2]Research Centre for Heart, Brain, Hormone and Healthy Ageing,
[3]Department of Medicine, University of Hong Kong,
Hong Kong, China

ABSTRACT

Adiponectin is an oligomeric adipokine possessing a myriad of beneficial activities against obesity-related medical complications. Abnormal circulating adiponectin levels are closely associated with the development of chronic liver diseases, including non-alcoholic hepatic steatosis, steatohepatitis, fibrosis and cirrhosis, as well as hepatocellular carcinoma. The hepatoprotective functions of adiponectin have been demonstrated by numerous clinical, pharmacological and genetic investigations. This review will provide an update on the pathophysiological role of adiponectin in liver and the molecular mechanisms underlying its metabolic and anti-inflammatory activities. The potentials for adiponectin-based therapeutic interventions for chronic liver diseases will also be discussed.

ABBREVIATIONS

ACC,	acetyl Coenzyme A carboxylase;
AKO,	adiponectin knockout mice;
AMPK,	AMP-activated protein kinase;
BMI,	body mass index;
CCD,	a conserved collagenous domain;

* Correspondence to Yu Wang, Department of Pharmacology and Pharmacy, University of Hong Kong, Rm 2-53, 21 Sassoon Road, Hong Kong. Tel: 852 28192864; Email: yuwanghk@hku.hk

CHC,	chronic hepatitis C;
CLD,	chronic liver disease;
CPT-1,	carnitine palmetoyl transferase 1;
CRP,	C-reactive protein;
CXCL8,	chemokine interleukin 8;
FAS,	fatty acid synthase;
gAd,	globular domain of adiponectin;
GGA1,	Golgi-localizing γ-adaptin ear homology ARF-binding protein-1;
GPAT,	glycerol-3-phosphate acyltransferase;
GPCR,	G-protein-coupled receptor;
HBV,	hepatitis B virus;
HCC,	hepatocellular carcinoma;
HCV,	hepatitis C virus;
HOMA-IR,	homeostasis assessment model-insulin resistance;
HSC,	hepatic stellate cells;
HVR,	hypervariable region;
HMW,	high molecular weight;
IL,	interleukin;
IL1RA,	interleukin-1-receptor antagonist;
LMW,	low molecular weight;
LPS,	lipopolysaccharide;
MCD,	methionine-and-choline deficient diet;
MMW,	middle molecular weight;
MRC,	mitochondria respiratory chain;
NAFLD,	nonalcoholic fatty liver disease;
NASH,	nonalcoholic steatohepatitis;
PDGF,	platelet-derived growth factor;
PGC1α,	PPARγ coactivator-1α;
PPARα,	peroxisome proliferators-activated receptor α;
ROS,	reactive oxygen species;
SOCS,	suppressor of cytokine signalling;
SREBP1c,	sterol regulatory element-binding protein 1c;
T2DM,	type 2 diabetes mellitus;
TGFβ1,	transforming growth factor β1;
TLR4,	Toll-like receptor 4;
TNF,	tumor necrosis factor;
SNPs,	single-nucleotide polymorphisms;
UCP2,	uncoupling protein 2.

Traditionally, adipose tissue (fat) is considered as an inert energy storage depot for excess calories. Evidence during the past two decades have revolutionized the classic view of adipose tissue, which is now recognized as an active endocrine organ to regulate metabolic homeostasis, cardiovascular physiology and immune functions in the body [1-3]. Adipose tissue produces a large amount of hormones, growth factors, cytokines, extracellular matrix proteins, and lipid chaperones, collectively called adipokines. Dysregulated production of adipokine and chronic inflammation in adipose tissue are characteristics of obesity and contribute to the associated pathological conditions, such as insulin resistance, cardiometabolic syndrome, and certain types of cancer [4-7]. Many adipokines, including leptin, adiponectin, lipocalin-2 and resistin, play active roles in modulating insulin sensitivity, inflammation and energy metabolism through local, peripheral and central mechanisms [8-12].

The close relationship between fat and liver tissues has been signified by the high prevalence of chronic liver disease associated with the growing pandemic of obesity during the past two decades [13, 14]. Under obese conditions, aberrant productions of free fatty acids, pro-inflammatory adipokines and chemokines from adipose tissue are particularly deleterious to the liver tissue, leading to the "adipocentric" concept of hepatic disorders [15, 16]. Numerous adipokines have been found to mediate the adipose-liver cross-talks [17-21]. This chapter will provide an update on how adiponectin, an anti-inflammatory adipokine, acts to protect liver functions, and the therapeutic potential of this protein in chronic liver diseases.

1. The Discovery and Diversified Biological Functions of Adiponectin

Adiponectin, also termed Acrp30, AdipoQ, apM1 or GBP28, was originally isolated and identified by four independent groups [22-25]. Adiponectin structurally belongs to the soluble defense collagen superfamily, sharing significant homology with collagen X, VIII and complement factor C1q [26]. The primary amino acid sequences of adiponectin are highly conserved, with over 80% identity across all the species cloned so far. Adiponectin is a relatively abundant adipokine, circulating in the blood stream at micrograms per millilitre range [27]. Unlike many other adipokines [e.g. tumor necrosis factor alpha (TNFα), interleukin 6 (IL6), leptin, lipocalin-2 and resistin] that are up-regulated in obese subjects, decreased circulating levels of adiponectin are found in many pathological conditions associated with obesity [28, 29]. In humans, adiponectin levels are reduced among men and obese subjects [30, 31], in patients with type 2 diabetes mellitus (T2DM) [32], hypertension [33] and coronary heart disease [34], and correlate with insulin sensitivity [35] and high-density lipoprotein cholesterol [36]. Ethnic differences of adiponectin expression have also been reported [37]. In obese rodents, reduced adiponectin levels are mainly due to the impaired synthesis in adipocytes, while the clearance of this protein in plasma is significantly prolonged [27].

This adipokine has attracted much attention because of its multiple beneficial biological activities. It is an insulin-sensitizing hormone critically involved in regulating glucose and lipid homeostasis [38-41]. Replenishment of adiponectin in rodents exerts protective functions against almost all the major obesity-related disorders. It can reduce body weight [42], prevent atherosclerosis and coronary heart disease [43, 44], ameliorate nonalcoholic fatty liver disease (NAFLD) and steatohepatitis (NASH) [45], attenuate obesity-related hypertension [46], suppress airway inflammation [47]. The anti-tumor properties of adiponectin have also been established in a number of life-style-related cancers [48, 49].

Adiponectin exerts its pleiotropic beneficial effects through direct actions on various target tissues and cells. In skeletal muscle, adiponectin reduces lipid accumulation by enhancing fatty acid β-oxidation and increases glucose uptake [40, 41]. In liver, adiponectin inhibits both gluconeogenesis and lipogenesis, resulting in alleviation of hyperglycaemia and hepatic steatosis [50-53]. In vascular system, adiponectin has been shown to be effective in preventing almost every pathological process involved in atherosclerotic plaque formation, including alleviation of endothelial dysfunction, inhibition of endothelium-leucocytes interaction, suppression of macrophage activation, macrophage-to-foam cell transformation and platelet aggregation [54, 55]. Adiponectin protects against myocardial infarction by inhibition of local TNFα production and cardiomyocyte apoptosis [43]. In addition, mounting evidence from both *in vitro* and *in vivo* studies have demonstrated adiponectin to be a negative regulator for cancer cell growth and survival [49, 56]. It can selectively bind to various carcinogenic growth factors and prevent their interactions with receptors [57]. In tumor tissues, adiponectin inhibits the growth and migration of vascular endothelial cells, prevents new blood vessel formation, and attenuates the growth of cancer cells [48, 58].

Once secreted, adiponectin may exert its effects through binding to a set of receptors, such as the adipoR1 and adipoR2 and/or T-cadherin [59, 60], and to various growth factors, cytokines and other interacting partners [57, 61, 62]. Although a number of downstream signalling cascades, including those mediated by AMP-activated kinase (AMPK) and peroxisome proliferators-activated receptor α (PPARα), are postulated to play a role in mediating the adiponectin signal, the detailed molecular events at the receptor as well as intracellular levels remain largely uncharacterized and somewhat controversial.

2. Structural Characteristics of Adiponectin: Dynamic Oligomerization and Posttranslational Modifications

Human adiponectin gene is located on chromosome 3q27 and encodes a 244 amino acids polypeptide comprising of an NH_2-terminal secretory signal sequence, followed by a hypervariable region (HVR), a conserved collagenous domain (CCD) with 22 Gly-X-Y repeats and a COOH-terminal C1q-like globular domain (gAd) [63]. High resolution three dimensional structures of full length adiponectin are not available, except that of the gAd of adiponectin, which has been resolved at a resolution of 2.1 Å as trimeric complexes [64]. This structure reveals an unexpected homology to TNF family of cytokines, both having the 10-β strand jellyroll folding topology to form a bell-shaped homotrimeric oligomer.

Endogenous adiponectin is presented as oligomeric complexes [65, 66]. The basic building block of the adiponectin complex is a tightly associated homotrimer, also called low molecular weight (LMW) oligomer, which is formed *via* hydrophobic interactions within the gAd and stabilized by the noncovalent interactions of the CCD. Freeze etch electron microscopy revealed that the trimer exhibits a ball- and stick-like structure containing a large globular sphere, an extended triple helix collagen stalk, and a smaller sphere on the opposite end of the stalk [67]. Two trimers self-associate to form a hexamer or middle molecular weight (MMW) oligomer. The assembly of hexameric adiponectin depends on the formation of a disulfide bond mediated by an NH_2-terminal conserved cysteine residue (C36 and C39 in human and murine adiponectin, respectively). Mutation of this cysteine to either alanine or serine leads to exclusive trimer formation, losing the capacity to further assemble into hexamer and the HMW oligomers [67, 68]. The architecture of hexameric adiponectin visualized by high resolution electron microscope shows two trimers lying adjacent to each other in parallel head-to-head fashion and is reminiscent of the letter Y [69]. Approximately 12-18 monomers assemble into a high molecular weight (HMW) multimeric complex of adiponectin [67, 69, 70]. The HMW oligomeric structures have been determined by single-particle analysis of electron micrographs [69]. *Pleomorphic* ensembles of collagen-like stretches of the trimers lead to a highly dynamic structure of HMW adiponectin, which could be classified into two major classes, the fan-shaped (class I) and bouquet-shaped (class II). In both of these conformations, the globular domains assume a variety of arrangements, covering an area of up to 4.9 × 105 $Å^2$ and up to 320 Å apart. The conformational flexibility of the HMW oligomer can allow it to access and cluster disparate target ligands or receptors, which may be necessary for activating cell signalling to elicit the remarkable functional diversity of adiponectin [63].

The highly dynamic characteristics of adiponectin oligomers confer this adipokine the capability to exert diversified biological functions. Different oligomers of adiponectin possess distinct biological activities in different targets. For example, the trimeric adiponectin is the most potent form involved in its insulin-sensitizing actions in skeletal muscle [67]. On the other hand, the HMW oligomeric complex of the adipokine is the major bioactive form responsible for inhibition of hepatic glucose production [71, 72], and for protection of endothelial cells from apoptosis [73]. In humans, the oligomeric complex distribution, but not the absolute amount of total adiponectin, determines insulin sensitivity [74]. Obese individuals have different distribution of adiponectin oligomers compared with lean control. Lower content of HMW adiponectin is closely associated with obesity-related metabolic complications [75]. The increases in the ratio of HMW versus total adiponectin, but not total adiponectin level *per se*, correlate well with improved insulin sensitivity during treatment with the insulin-sensitizing drug thiazolidinediones in both diabetic mice and patients with T2DM [75]. On the other hand, weight reduction by either calorie restriction or gastric bypass surgery results in a selective elevation of the HMW adiponectin, but not the trimeric and hexameric complexes [76-78]. In line with these data from the epidemiological studies, there is also genetic evidence supporting the role of HMW adiponectin as a major insulin-sensitizing isoform in human. Two rare mutations (G84R and G90S) located within the collagenous domain are closely associated with insulin resistance and T2DM [79]. Interestingly, subjects with either of these mutations have extremely low level of HMW adiponectin. Moreover, recombinant adiponectin with either of these mutations expressed in NIH-3T3 fibroblasts fail to form HMW oligomers. These data suggest that the beneficial effects of adiponectin in humans might be mediated primarily by its HMW isoform, and the deficiency of this oligomer is an important etiological factor that links obesity with its medical complications.

Analysis of endogenous adiponectin by two-dimensional gel electrophoresis reveals that this adipokine is heavily glycosylated to form a cluster of heterogeneous "spots" with different isoelectric points and molecular masses [51, 63]. Studies performed on murine, bovine and human adiponectin have identified several conserved prolines in CCD that undergo hydroxylation, and five conserved lysines (one in HVR and four in CCD) that are hydroxylated and subsequently glycosylated by a glucosyl α [1-2] galactosyl group [51, 80, 81]. These glucosylgalactosyl hydroxylysine residues participate in intra- and intermolecular interactions and appear to be important for the formation of the HMW oligomeric complex of adiponectin. Depletion of hydroxylation and glycosylation by mutation of the lysines to arginines abrogates the intracellular assembly of the HMW oligomers [72, 80], and abolishes the insulin sensitizing and receptor binding activity of adiponectin [82, 83].

The precise enzymes responsible for these posttranslational modifications remain elusive at this stage. A family of lysyl hydroxylases (EC 1.14.11.4), consisting of lysyl hydroxylase-1, -2a, -2b, and -3, has been shown to catalyze the conversion of lysine into hydroxylysine [84]. Pharmacological inhibition of these enzymes results in a reduced glycosylation of adiponectin secreted from rat adipocytes [72]. Among these four members, lysyl hydroxylase-3 is the most likely candidate involved in the modification of adiponectin. In addition to its lysyl hydroxylase activity, lysyl hydroxylase-3 also possesses a low level of glucosyltransferase and galactosyltransferase activities, suggesting that this enzyme alone is sufficient for catalyzing lysine hydroxylation and its further attachment with a glucosylgalactosyl group [85].

Another type of posttranslational modification on adiponectin is succination, in which a thioether bond is formed between the Krebs cycle intermediate fumarate and cysteine residues to generate S-(2-succinyl)cysteine [86]. Succination at the NH_2-terminal conserved cysteine residue blocks the oligomerization of adiponectin by inhibition of disulphide bond formation. The extent of succination on intracellular adiponectin protein is elevated in adipose tissues of obese and diabetic mice, suggesting that this modification may contribute to impaired adiponectin secretion in obesity and related disorders [86]. Adiponectin is also modified by a α2,8-linked disialic acid moiety (Neu5Acα2→8Neu5Acα2→3Gal) [87]. This modification occurs on O-linked glycans situated at Thr residues within HVR [88]. Loss of sialylation has no effect on adiponectin oligomerization, secretion or activity, but causes its rapid clearance from the circulation, suggesting a role for this modification in determining the half-life of circulating adiponectin.

3. Adiponectin and Chronic Liver Diseases

The global epidemic of obesity has significantly increased the prevalence of chronic liver diseases (CLD), including non-alcoholic fatty liver disease, chronic infection with hepatitis C virus (HCV) and hepatic cell carcinoma [13, 89, 90]. Under obese condition, excess fat deposition, particularly around abdomen, impacts on liver functions by altering hepatic lipid metabolism and promoting imbalanced adipokine and/or cytokine production [91-93]. For example, leptin can elicit anti-steatotic effects in patients with severe lipodystrophy; however, the elevated leptin levels in obese condition promote liver fibrogenesis [94-96]. TNFα, a pro-inflammatory adipokine, interferes with insulin signalling and favours steatosis, and plays a casual role in the pathogenesis of liver injury [97-99]. Neutralization of TNFα activity improves fatty liver disease in animals [100]. On the other hand, nutritional steatohepatitis can still be produced experimentally in both TNFα and TNFα type 1 receptor knockout mice [101, 102]. In contrast to leptin and TNFα, adiponectin is more closely implicated in the pathogenesis of CLD. Low serum adiponectin concentrations have been found in patients with liver dysfunctions. There are also countless evidences to support the protective role of adiponectin against liver injuries. In the following sections, the detailed relationship between adiponectin and several types of CLD will be discussed.

3.1. Non-Alcoholic Fatty Liver Disease

Non-alcoholic fatty liver disease (NAFLD) is the most common type of chronic liver injury in many countries [103, 104]. NAFLD includes a spectrum of diseases ranging from simple steatosis, nonalcoholic steatohepatitis (NASH) to fibrosis, cirrhosis and hepatocellular carcinoma [105]. Obesity, especially visceral obesity, is frequently associated with NAFLD and their coexistence in the same individual increases the likelihood of having more advanced forms of liver disease [106, 107]. The overall prevalence of NAFLD is between 20% and 40% in general adult population, and is much higher in obese people (60-95%) than that in lean subjects (16%) [108, 109]. Mortality in patients with NAFLD is significantly higher than that in the age and gender-matched general population [110]. Disease progression to NASH, cirrhosis and life-threatening advanced liver disease appears to be slow. In many cases of NAFLD, risks of developing metabolic and cardiovascular morbidities are much higher than the hepatic diseases [111-113]. NAFLD significantly increases the risk of diabetes, and is a better predictor of the development of metabolic disorders than obesity itself [114]. Recent studies have reported an association of NAFLD with multiple classical and non-classical risk factors for cardiovascular diseases [111, 115]. NAFLD predicts future cardiovascular events independently of other prognostic factors, including the component of metabolic syndrome, and thus should be considered beyond a chronic hepatic disease.

Numerous epidemiological investigations have identified hypoadiponectinemia as an independent risk factor for NAFLD in diversed ethnic groups [116]. Plasma adiponectin level is an independent indicator of liver fat accumulation [117]. A negative association between serum adiponectin and liver enzyme levels has been shown in obese subjects [118, 119]. Systemic adiponectin is about 20% to 60% lower in NAFLD than healthy controls [120, 121]. It is suggested that serum adiponectin may be useful to estimate the severity of liver damage in both obese and non-obese patients with NAFLD [122]. Circulating adiponectin levels are inversely associated with the degree of hepatic steatosis, necroinflammation and fibrosis, and this correlation is independent of insulin resistance and body mass index (BMI) [119, 123]. During the development of NAFLD, from simple steatosis to NASH, adiponectin expression is decreased by 20%~40% [123, 124]. Compared with healthy controls, adiponectin levels are lower by more than 50% in NASH patients [123]. Shimada et al reported that 90% of patients with early-stage NASH can be predicted by a combined evaluation of the serum adiponectin level, homeostasis assessment model-insulin resistance (HOMA-IR) score, and serum type IV collagen 7S level [124]. In patients with T2DM, plasma adiponectin concentrations are inversely related to hepatic fat content [125]. In metabolically healthy obese population, despite excessive weight and body fat, individuals that are insulin sensitive also show adiponectin levels similar to those in normal BMI subjects [126]. Genetic variations in the adiponectin gene can affect its circulating level [79]. The prevalence of single-nucleotide polymorphisms (SNPs) 45GT is found to be higher in the NAFLD patient group compared with healthy subjects [127, 128]. SNPs in adipoR1 and adipoR2 also contribute to variations in hepatic fat accumulation in humans [129, 130].

3.2. Liver Fibrosis and Cirrhosis

Liver fibrosis is the hallmark of severity in CLD, characterized by the accumulation of extracellular matrix proteins. It is the consequence of chronic or repeated liver injury caused by hepatotoxic agents such as alcohol and viruses, as well as immune and metabolic disorders. Insulin resistance and T2DM have been identified as risk factors for liver fibrosis [131, 132]. Other parameters of metabolic syndrome, such as high blood pressure and hyperinsulinemia, are also associated with fibrosis [133, 134]. Low adiponectin levels are found in morbidly obese patients with liver fibrosis and correlated with more severe liver histology [122, 135, 136]. Moreover, adiponectin SNPs are associated with the progression of liver fibrosis [137, 138]. Interestingly, both adiponectin and its receptors have been shown to be decreased in NASH livers compared with simple steatosis, and to be associated with higher inflammation and fibrosis [139, 140]. On the other hand, contradictory results have been obtained showing that hypoadiponectinemia in NAFLD only correlates with hepatic insulin resistance and fat content, but not to liver disease severity [121]. During the development of liver fibrosis, the serum level of total adiponectin significantly increases [141]. Despite these discrepancies, the anti-inflammatory properties of adiponectin support that this protein protects the steatotic liver from progression to fibrosis and that diminished adiponectin levels associated with obesity and metabolic syndrome may contribute to the aggressive fibrotic response to local inflammation within the liver.

Liver is one of the major organs for adiponectin extraction and catabolism. There is a significant amount of adiponectin exchange in hepatic vascular bed [27]. In both human subjects and animal models, bile duct ligation causes rapid increase of adiponectin levels, indicating a potential biliary route of adiponectin clearance [142]. In fact, adiponectin could serve as a novel marker for cholestasis in liver cirrhosis, an advanced stage of fibrosis [143, 144]. It has long been known that the progression of chronic liver disease to cirrhosis is accompanied by the development of insulin resistance [145]. Patients with cirrhosis have elevated levels of adiponectin, as well as fasting glucose, insulin and catecholamines [146, 147]. However, the plasma concentrations of adiponectin in cirrhosis do not correlate with insulin resistance or parameters of body composition and energy metabolism [148]. The increased adiponectin levels in cirrhosis correlate exclusively with reduced liver function and altered hepatic hemodynamics [146]. The metabolic consequences of raised adiponectin in patients with established liver cirrhosis warrant further study.

In summary, while the current evidences demonstrate a close link between altered adiponectin levels and liver dysfunction, studies in larger cohorts of patients with detailed assessment of metabolic and other potential confounding factors should provide further insights to dissect the associations of adiponectin with NAFLD, especially in the presence of systemic insulin resistance and overt diabetes. The effects of this adipokine on liver histological lesions and various hepatic cells should be evaluated in prospective, carefully designed studies. Clearly, when interpreting adiponectin levels in patients with NAFLD, one has to take into account the stage of the disease: if characteristically low in early stages, it will increase with the progression of the disease to actually be elevated in patients with cirrhosis.

3.3. Chronic Hepatitis and Hepatocellular Carcinoma

NAFLD with obesity predisposes an individual to viral liver diseases, such as chronic hepatitis C (CHC) infection and the subsequent fibrotic progression [112, 149]. Hepatic steatosis and fibrosis are common histological findings in patients with CHC [150]. The alterations of plasma adiponectin are found to be gender- and/or genotype-specific, and in general associated with grades of steatosis and HOMA-IR in chronic HCV patients [151-154]. Patients with hepatitis C virus (HCV) genotype 4 infection with steatosis showed lower serum adiponectin levels and higher levels of leptin [151]. In multivariate analysis, male gender, insulin resistance, high HCV viral load and genotype 2 are associated with decreased serum adiponectin level [155]. In patients infected with HCV genotype 3, adiponectin levels were reduced, independently of the extent of steatosis. Adiponectin levels are increased after successful antiviral treatment [156]. Low adiponectin levels are associated with poor response to antiviral therapy [157]. Patients with HCV genotype 1 infection had significantly higher insulin resistance than those with genotype 2 infection and lower serum adiponectin concentration [158]. Patients with none or mild fibrosis show similar serum adiponectin levels with those in healthy controls, whereas hyperadiponectinemia is found in those with moderate to severe stages of fibrosis [159, 160]. There is no adiponectin cutoff level that has reasonable sensitivity/specificity for predicting hepatic fibrosis stage. On the other hand, adiponectin could be used as a predictor for antiviral response to reflect the improvement in hepatic inflammation [161]. The contribution of HCV to metabolic disorders, including insulin resistance, impaired glucose tolerance, type 2 diabetes and abnormal lipid metabolism, has been suggested [162]. HCV may promote insulin resistance through direct viral-dependent mechanisms or due to activation of the inflammatory response, resulting in increased production of TNFα. Whether HCV infection directly influences adiponectin synthesis merits further investigation. Hepatitis B virus (HBV) infection does not affect adiponectin levels. Serum adiponectin level correlates with the progression of HBV-related liver diseases but not with the development of HBeAg seroconversion [163, 164]. A marked decline in serum adiponectin is found after antiviral therapy and associated with fibrosis reduction [165]. Taken together, although adiponectin is participated in the pathogenesis of chronic hepatitis, many aspects need to be extensively investigated and elucidated, such as the complex role of this adipokine in the interactions among insulin resistance, hepatic necroinflammation and chronic HCV infection.

The potential roles of hypoadiponectinemia in hepatocellular carcinoma (HCC) have been documented [116, 166], which underscore the importance of abnormal adiponectin signalling in the development and progression of NAFLD from simple steatosis to its complications. A strong epidemiological association between obesity, insulin resistance, diabetes and the risk for HCC has been established [167-169]. Adiponectin together with insulin resistance is implicated in hepatocarcinogenesis [170-174]. Insulin resistance and hypoadiponectinemia may be implicated in hepatocarcinogenesis by promoting hepatic fibrosis [171, 175]. Obesity and metabolic syndrome are also recognized risk factors for hepatocellular carcinoma (HCC) of patients with CHC, in which decreased production of adiponectin appears to play a role. Serum levels of leptin, insulin, HOMA-IR and BMI were significantly higher, whereas serum adiponectin was significantly lower in patients with HCC at the time of tumor detection, at more than five years after a sustained virological response to interferon treatment [175]. On the other hand, in patients with compensated HCV cirrhosis and persistent infection, insulin resistance but not serum levels of adiponectin and leptin predict the occurrence of HCC [170]. In another large-scale retrospective cohort study, unexpectedly, CHC patients with high serum adiponectin levels showed a higher risk of liver cancer development [174]. T-cadherin, a receptor of adiponectin, was found to be over-expressed on intratumoral capillary endothelial cells in a less differentiated HCC region [176].

3.4. Hepatoprotective Functions of Adiponectin

3.4.1. Animal Models

The hepatoprotective and antifibrogenic effects of adiponectin have been demonstrated in animal models by numerous pharmacological, genetic, gain- and loss-of-function studies [123, 177]. Xu et al first reported that in animal models of both alcoholic and nonalcoholic steatohepatitis, administration of exogenous adiponectin could reduce hepatomegaly, deplete lipid accumulation, quench hepatic inflammation and decrease hepatic expression of TNFα [52]. Later on, pharmacological approaches to treat various mice models with adiponectin revealed that this protein can elicit potent protective activities against other forms of liver injuries, induced by carbon tetrachloride, lipopolysaccharide (LPS)/D-galactosamine, chemical compounds, bile duct ligation and nutritionally deficient diet [178-182]. Hepatic resistance to adiponectin has been described in animal models of nutritional steatohepatitis caused by methionine-and-choline deficient diet (MCD) [183]. To explore the potential role of adiponectin in NAFLD/NASH pathogenesis, mice lacking adiponectin (adiponectin knockout mice, AKO) have been widely used. There mice are more prone to high fat diet- and genetic obesity-induced NAFLD and NASH [184]. Histologically, steatosis appears more pronounced in AKO livers but hepatic triglyceride content is similar to that of wild type mice fed a high fat diet [185]. Instead, the intermediate lipid metabolites are significantly elevated in AKO mice livers [186]. Lack of adiponectin also enhances LPS/D-galactosamine-induced liver injury [181]. Compared to wild type littermates, the anti-inflammatory and the mitochondria-stimulatory activity of rosiglitazone, a PPARγ agonist, are largely attenuated in AKO mice. Prolonged exposure to high fat diet induces sustained perisinusoidal fibrosis and much higher expression levels of TNFα in the livers of AKO mice [185]. In response to a choline-deficient L-amino acid-defined diet, AKO mice undergo an enhanced progression to NASH and fibrosis, and develop hepatic tumors with a higher incidence when compared to wild type mice [187]. Adiponectin deficiency is associated with elevated oxidative stress markers, TBARS and 8-OHdG, in liver tissues. In addition, adiponectin deficiency enhances liver tumor formation induced by azoxymethane [188]. These results collectively indicate that lack of adiponectin enhances the progression of hepatic steatosis, steatohepatitis, fibrosis, and the development of hepatic tumors, which is in line with the clinical evidence suggesting that hypoadiponectinemia in obesity contribute to the high prevalence of CLD.

3.4.2. Cellular Studies

Inflammatory cytokines are key mediators of hepatic inflammation, cell death, fibrosis, as well as regeneration after massive or focal liver injury [93]. Adiponectin levels are negatively associated with mediators of inflammation, including IL6 and C-reactive protein

(CRP), but positively related to anti-inflammatory cytokine IL10 [36]. Adiponectin suppresses TNFα functions via inhibition of its expression and antagonizing its activities [52, 182, 184, 189]. In liver, cytokines such as IL6 and TNFα are mainly produced from Kupffer cells and hepatic stellate cells (HSC) [123, 190, 191]. Adiponectin ameliorates NASH and liver fibrosis through suppressing the activation of Kupffer cells and HSCs. In addition, adiponectin modulates the inflammatory response in endothelial cells by inhibiting NF-κB activation and the release of TNFα [192]. Adiponectin can also induce the expression of anti-inflammation cytokine interleukin-1-receptor antagonist (IL1RA) [193, 194]. In porcine blood-derived macrophages, adiponectin suppresses both TNFα and IL6 production stimulated by LPS and induces IL10 expression. The attenuation of proinflammatory cytokine production by adiponectin is mediated in part by attenuating the translocation of NFκB to the nucleus [195]. The anti-inflammatory effects of adiponectin in macrophages may involve Toll-like receptor 4 (TLR4) signalling pathway. However, the mechanisms by which adiponectin suppresses TLR4-mediated responses are not well understood [196]. Transformation of HSCs into myofibroblasts is the key step initiating the fibrotic process during liver injury [91, 96]. The activated HSCs increase the accumulation of extracellular matrix. Both adiponectin receptors, adipoR1 and adipoR2, are expressed in HSCs. Adiponectin treatment inhibits platelet-derived growth factor (PDGF)-stimulated proliferation and migration of human HSCs, and reduces the secretion of monocyte chemoattractant protein-1 through AMPK-dependent mechanisms [96, 197, 198]. Additionally, adiponectin also regulates hepatic expression of transforming growth factor β1 (TGFβ1), a pro-fibrotic factor involved in HSC activation [179, 199]. Inhibition of adipoR2 expression by short hairpin RNAi-expressing adenovirus can induce TGFβ1 expression, and overexpression of adipoR2 diminishes TGFβ1 mRNA level. Adiponectin upregulated the chemokine interleukin 8 (CXCL8), an anti-apoptotic protein, *via* adipoR1 and NFκB dependent pathways in primary human hepatocytes, thereby may prevent hepatocyte apoptosis [200]. It further antagonizes hepatocyte death by blocking fatty acid-induced activation of c-Jun NH2 terminal kinase and by inhibiting fatty acid mediated upregulation of CD95 [201].

3.4.3. Molecular Mechanisms Underlying the Hepatoprotective Functions of Adiponectin

Adiponectin regulates hepatic lipid/glucose homeostasis and elicits insulin sensitizing activities through activation of AMPK and PPARα, two important pathways controlling energy partition and metabolism [177, 202]. AMPK is a fuel-sensing enzyme that stimulates adenosine triphosphate (ATP)-generating catabolic pathways, such as glycolysis and lipid oxidation, and turns off energy-consuming processes, such as glycogen, lipid and protein synthesis, to restore energy balance. The precise mechanisms whereby adiponectin activates AMPK through its receptors remain to be determined. APPL1, an adaptor protein containing pleckstrin homology domain, phosphotyrosine binding domain and leucine zipper motif, appears to be a key signaling molecule that couples adiponectin receptors and its downstream AMPK activation [203, 204]. Adiponectin enhances the binding of APPL1 to both adipoR1 and adipoR2, and these interactions are essential for subsequent phosphorylation and activation of AMPK, which phosphorylates acetyl Coenzyme A carboxylase (ACC) and attenuates its activity. Inhibition of ACC reduces lipid synthesis and enhances fatty acid oxidation by blocking the production of malonyl-CoA, an allosteric inhibitor of carnitine palmetoyl transferase 1 (CPT-1), the rate-limiting enzyme in fatty acid oxidation. In addition, AMPK downregulates the expression of sterol regulatory element-binding protein 1c (SREBP1c), a transcription factor regulating cholesterol and lipid synthesis. Reduction of SREBP1c results in downregulation of genes involved in lipogenesis, including ACC, fatty acid synthase (FAS), and glycerol-3-phosphate acyltransferase (GPAT) [205-207]. PPARα is a transcription factor controlling the transcription of a panel of genes encoding fatty acid oxidation enzymes, such as FATP, acyl-CoA oxidase and long chain acyl-CoA synthetase. Adiponectin stimulates PPARα activity possibly through PPARγ coactivator-1α (PGC1α) [208]. PPARα upregulates CPT-1, stimulates β-oxidation, reduces lipid synthesis and thereby prevents excess triglyceride storage [209]. Together, activation of these signaling pathways by adiponectin leads to enhanced fat oxidation, reduced lipid synthesis and prevention of hepatic steatosis [177].

Hepatic lipid overload increases the formation of reactive oxygen species (ROS) and lipid peroxidation products, which in turn promote the generation of proinflammatory cytokines and facilitate the progression of liver injury [210]. Elevated cytochrome P450 2E1 levels are associated with the enhanced oxidative stress in the liver tissues of AKO mice under nutritional deficient diet-induced NASH condition, and could be down-regulated by replacement with adiponectin [187]. Serum TNFα levels are significantly augmented in AKO mice [211, 212]. The anti-inflammatory activities of adiponectin are in part mediated by suppressing TNFα production [124,125]. Administration of adiponectin in sinusoidal endothelial cell cultures induces a decreased production of TNFα [131]. Aldehyde oxidase 1 (AOX1) is a xenobiotic metabolizing protein and an important source of ROS and is reduced in hepatocytes by adiponectin *via* activation of PPARα [213, 214]. Adiponectin also increases ROS detoxifying enzymes and adipoR2 is involved in the induction of superoxide dismutase 1 and catalase [215].

Mitochondrial dysfunction represents a unique characteristic of the liver tissues in mice without adiponectin [184]. The hepatic mitochondria exhibit ultrastructural lesions and decreased activity of the respiratory chain complexes. In this condition, the decreased respiratory chain activity results in accumulation of ROS that oxidize fat deposits to form lipid peroxidation products, which in turn, cause steatohepatitis, necrosis, inflammation and fibrosis. Adiponectin treatment restores the activities of the hepatic mitochondrial respiratory chain complexes (MRC) and decreases the levels of lipid peroxidation products through upregulating uncoupling protein 2 (UCP2), a mitochondria inner membrane transporter [184, 186]. In UCP2 deficient mice, the effects of adiponectin on MRC activities are dramatically attenuated, suggesting that the increased UCP2 expression might be obligatory for adiponectin to elicit its activities on mitochondria functions. UCP2 is required for adiponectin-mediated AMPK activation, fatty acid oxidation and anti-oxidative stress activities [186]. It is known that UCP2 inhibits excess ROS production from mitochondria [216]. It can also prevent the production of pro-inflammatory cytokines from both macrophage and Kupffer cells [217]. Although these evidence suggest the existence of a reciprocal relationship between UCP2 and adiponectin, the detailed signalling mechanisms underlying adiponectin-induced UCP2 expression are largely uncharacterized.

Circulating levels of adiponectin (μg/ml range) by far exceed the concentrations commonly required for receptor-dependent signalling, which indicates a receptor-independent function of adiponectin. In fact, binding to growth factors like PDGF, extracellular matrix proteins, low density lipoprotein and opsonization of apoptotic cells for phagocytosis have been described [57, 218-220]. Adiponectin inhibits the effects of TGFβ_1 and PDGF on HSC proliferation/migration, suppresses the expression of connective tissue growth factor, and nuclear translocation of Smad2 [179]. AMPK mediates the suppressive effects of adiponectin on proliferation of activated HSCs [198]. Leptin is a profibrotic adipokine. Activation of AMPK disrupts leptin-mediated hepatic fibrosis by upregulating suppressor of cytokine signaling (SOCS3), blocking TIMP-1 expression and stimulating MMP-1 activity in mice HSCs [133]. In HCC cell lines HepG2 and Huh7, adiponectin could antagonize the oncogenic actions of leptin by reducing Stat3 and Akt phosphorylation. Adiponectin also increases SOCS3, a physiologic negative regulator of leptin signal transduction [221]. In addition, adiponectin modulates other pathways, such as C-jun N-terminal kinase and mammalian target of rapamycin to inhibit HCC cell growth/survival [172, 173] .

Collectively, the above studies performed in animal models as well as in isolated cells suggest that adiponectin is a potent hepatoprotecive agent that can elicit a wide range of beneficial functions to regulate insulin sensitivity, energy metabolism, mitochondrial activity, oxidative and inflammatory responses, fibrogenesis and carcinogenesis in liver.

4. HMW Adiponectin Is the Major Bioactive Form in Liver

Evidence from both *in vitro* and animal-based studies supports the role of the HMW oligomer as the major active form in mediating the multiple beneficial actions of adiponectin in liver. The formation of the HMW oligomers is obligatory to mediate the insulin-sensitizing effects of adiponectin on suppression of hepatic gluconeogenesis in primary rat hepatocytes [51, 82]. Intravenous injection of the HMW adiponectin, but not the hexameric adiponectin, led to a dose-dependent decrease in serum glucose levels and a marked activation of AMPK in the liver [75], while chronic infusion with this protein leads to prolonged alleviation of hyperglycaemia and insulin resistance in *db/db* diabetic mice [72]. Recombinant adiponectin produced from mammalian cells, which can form the HMW oligomers, potently decreased hyperglycaemia in diabetic mice through inhibition of hepatic glucose production [50, 72]. On the other hand, bacterially generated full-length adiponectin, which lacks the capacity to form the HMW adiponectin, was almost inactive [81]. Intravenous injection of the HMW adiponectin, but not the hexameric adiponectin, leads to a dose-dependent decrease in serum glucose levels [75]. These animal-based evidences are consistent with the clinical observations showing that the ratio of HMW/total adiponectin correlates closely with hepatic insulin sensitivity [63]. Notably, treatment with thiazolidinediones has been shown to cause a selective elevation of the HMW oligomeric adiponectin [75]. In addition to the hepatic insulin-sensitizing activity, the HMW adiponectin has also been suggested to be the most potent isoform for alleviation of fatty liver disease in high fat diet-induced obese mice [81], and inhibition of apolipoprotein B and E release from human hepatocytes [222]. An independent inverse association exists between liver enzymes and HMW adiponectin [223]. HMW adiponectin dose-dependently suppressed growth factor-induced hepatic stellate cell proliferation [198]. Together, these data suggest that the beneficial effects of adiponectin in the hepatic tissue are mediated predominantly by its HMW form.

The structural properties of the HMW adiponectin remain poorly characterized at this stage, due to the heterogeneous nature of this oligomeric form [69]. It is important to note that the compositions of human HMW adiponectin oligomers differ from those observed in rodents and bovine. Analysis of adiponectin oligomers by non-denaturing and non-heating gel electrophoresis shows that the HMW oligomers of human serum adiponectin are composed of multiple species, ranging from 18-30 mers or even larger [79, 80]. A significant portion of *de novo* synthesized adiponectin is not secreted and retained in adipocytes. Treatment of cells with the reducing reagent such as β-mercaptoethanol stimulates the release of intracellular adiponectin by seven to eight folds. Secretion of the adiponectin oligomers is tightly controlled by a pair of molecular chaperones in endoplasmic reticulum, including ERp44 and Ero1-Lα [224-227]. ERp44 inhibits the secretion of adiponectin oligomers through a thiol-mediated retention. By contrast, Ero1-Lα releases the HMW oligomeric adiponectin trapped by ERp44. Because maturation of adiponectin oligomers needs extensive posttranslational modifications, this thio-mediated retention by ERp44 within the ER may prolong the resident time for this adipokine in the secretory pathway and thus provide a better chance for proper folding into the higher-order complexes. The role of ERp44 and Ero1-Lα in regulating the secretion of adiponectin oligomeric complexes is also corroborated by the observation that changes in expression of these two proteins are closely associated with alterations in the oligomeric complex distribution of adiponectin under various pathophysiological conditions [225]. In *ob/ob* obese mice, the decreased expression of ERp44 and Ero1-Lα in adipose tissue is associated with the reduced ratio of HMW to total adiponectin in the circulation. Liu and colleagues demonstrated that DsbA-L, another chaperone and a catalyst for disulfide bond formation, interacts with adiponectin and participates in HMW formation [228]. The interaction between DsbA-L and adiponectin may induce a conformational change in adiponectin and thus favors a disulfide bond exchange reaction for the assembly and secretion of adiponectin [229]. The PPARγ agonists thiazolidinediones selectively enhance the secretion of HMW oligomeric adiponectin through transcriptional upregulation of these critical chaperones ERp44, Ero1-Lα and DsbA-L [224]. Moreover, regulation of the HMW oligomeric adiponectin by the NAD-dependent deacetylase SIRT1 is also mediated by Ero1-Lα [230]. The SIRT1 activator resveratrol suppresses the secretion of HMW oligomeric adiponectin through deacetylation and inhibition of PPARγ, resulting in the downregulation of the PPARγ responsive gene Ero1-Lα. On the other hand, glucose and lactate selectively enhance the secretion of HMW oligomeric adiponectin through inhibition of SIRT1, thus leading to activation of PPARγ and upregulation of Ero1-Lα. The release of adiponectin from endoplasmic reticulum into the extracellular medium is mediated by a trafficking vehicle containing Golgi-localizing γ-adaptin ear homology ARF-binding protein-1 (GGA1) [231]. Overexpression of wild type GGA1 enhances adiponectin secretion from adipocytes, whereas overexpression of GGA1 dominant-interfering mutant causes a marked inhibition.

In summary, our current understanding of HMW adiponectin formation still has many gaps. It is known that once released from adipocytes, the different adiponectin oligomeric forms are not interchangeable [68]. Considering the versatile role of this adipokine in protecting liver functions, it is crucial to elucidate the detailed molecular mechanisms that govern the synthesis, assembly and secretion of HMW adiponectin at the level of adipocytes.

5. Elevation of Adiponectin Production: A Therapeutic Strategy for CLD

To date, there have been few specific and effective drug treatments for NAFLD and NASH. Early diagnosis and management of the underlying condition remains to be the mainstay of treatment. Weight loss and exercise have been shown to improve insulin resistance and to ameliorate fatty liver disease [108, 232]. These “life-style adjustment” or anti-obesity measures (including bariatric surgery) reduces steatosis, inflammation and hepatic fibrosis [233-235]. Increased adiponectin concentrations may partly contribute to the metabolic improvements observed in these treatments. As insulin resistance remains the pivotal etiology responsible for CLD, insulin sensitizers such as metformin and glitazones are currently being evaluated as the treatment of choice. Long-term metformin treatment significantly reduced alanine aminotransferase levels and marginally elevated serum adiponectin levels in NAFLD patients with steatohepatitis [236]. A greater benefit with a combination of atorvastatin and metformin in improving liver injury in T2DM patients with hyperlipidaemia has been shown to be associated with a significant elevation of plasma adiponectin [237]. Treatment with glitazones can effectively improve insulin sensitivity and decrease hepatic fat content [238]. Circulating levels of adiponectin are significantly elevated by both pioglitazone and rosiglitazone, supporting a protective role for adiponectin in the treatment of steatosis hepatitis.

Pharmacological intervention to elevate adiponectin production might hold promise for the treatment and/or prevention of NAFLD and NASH. For example, two structurally related natural compounds (astragaloside II and isoastragaloside I) isolated from the medicinal herb Radix Astragali selectively increase adiponectin secretion in primary adipocytes, and raise circulating concentrations of adiponectin in mice with dietary or genetic obesity. Furthermore, chronic administration of astragaloside II and isoastragaloside I in both dietary and genetic obese mice significantly elevated serum levels of total adiponectin and selectively increased the composition of its high molecular weight oligomeric complex [239]. These changes are associated with an alleviation of hyperglycemia, glucose intolerance and insulin resistance. These results suggest that pharmacological elevation of circulating adiponectin alone is sufficient to ameliorate insulin resistance and diabetes. The two natural compounds might also provide the lead as a novel class of therapeutics for obesity-related diseases, such as NAFLD, and could potentially be used to halt the progression of liver fibrosis.

Conclusion

Adiponectin is an abundant adipokine crucially involved in the cross-talks between adipocytes and hepatocytes. The importance of adiponectin in CLD has been increasingly recognized. With the well established anti-inflammatory and insulin sensitizing properties, adiponectin-based theraputics for NAFLD, the most common causes of chronic liver disease worldwide, represent a promising area for further investigation.

ACKNOWLEDGMENT

Research in the author's laboratory is supported by grants from Research Grants Council of Hong Kong (Project number 777908) and the collaborative research fund (HKU3/CRF/09) from the Research Grants Council of Hong Kong.

REFERENCES

[1] Galic S, Oakhill JS, Steinberg GR. Adipose tissue as an endocrine organ. *Mol Cell Endocrinol* 2011;316:129-139.
[2] Rasouli N, Kern PA. Adipocytokines and the metabolic complications of obesity. *J Clin Endocrinol Metab* 2008;93:S64-73.
[3] Ouchi N, Parker JL, Lugus JJ, Walsh K. Adipokines in inflammation and metabolic disease. *Nat Rev Immunol* 2011;11:85-97.
[4] Deng Y, Scherer PE. Adipokines as novel biomarkers and regulators of the metabolic syndrome. *Ann N Y Acad Sci* 2011;1212:E1-E19.
[5] Grossmann ME, Ray A, Nkhata KJ, Malakhov DA, Rogozina OP, Dogan S, Cleary MP. Obesity and breast cancer: status of leptin and adiponectin in pathological processes. *Cancer Metastasis Rev* 2010;29:641-653.
[6] Hui X, Lam KS, Vanhoutte PM, Xu A. Adiponectin and Cardiovascular Health: an Update. *Br J Pharmacol* 2011.
[7] Vlasova M, Purhonen AK, Jarvelin MR, Rodilla E, Pascual J, Herzig KH. Role of adipokines in obesity-associated hypertension. *Acta Physiol (Oxf)* 2010;200:107-127.
[8] Wauman J, Tavernier J. Leptin receptor signaling: pathways to leptin resistance. *Front Biosci* 2011;17:2771-2793.
[9] Wang Y. Small lipid-binding proteins in regulating endothelial and vascular functions: focusing on adipocyte fatty acid binding protein and lipocalin-2. *Br J Pharmacol* 2011.
[10] Schwartz DR, Lazar MA. Human resistin: found in translation from mouse to man. *Trends Endocrinol Metab* 2011;22:259-265.
[11] Ziemke F, Mantzoros CS. Adiponectin in insulin resistance: lessons from translational research. *Am J Clin Nutr* 2010;91:258S-261S.
[12] Antuna-Puente B, Feve B, Fellahi S, Bastard JP. Adipokines: the missing link between insulin resistance and obesity. *Diabetes Metab* 2008;34:2-11.
[13] Vernon G, Baranova A, Younossi ZM. Systematic review: the epidemiology and natural history of non-alcoholic fatty liver disease and non-alcoholic steatohepatitis in adults. *Aliment Pharmacol Ther* 2011;34:274-285.
[14] Fan JG, Saibara T, Chitturi S, Kim BI, Sung JJ, Chutaputti A. What are the risk factors and settings for non-alcoholic fatty liver disease in Asia-Pacific? *J Gastroenterol Hepatol* 2007;22:794-800.
[15] Parola M, Marra F. Adipokines and redox signaling: impact on fatty liver disease. *Antioxid Redox Signal* 2011;15:461-483.
[16] MacLaren R, Cui W, Cianflone K. Adipokines and the immune system: an adipocentric view. *Adv Exp Med Biol* 2008;632:1-21.

[17] Polyzos SA, Kountouras J, Zavos C, Tsiaousi E. The role of adiponectin in the pathogenesis and treatment of non-alcoholic fatty liver disease. *Diabetes Obes Metab* 2010;12:365-383.

[18] Wree A, Kahraman A, Gerken G, Canbay A. Obesity affects the liver - the link between adipocytes and hepatocytes. *Digestion* 2011;83:124-133.

[19] Bethanis SK, Theocharis SE. Leptin in the field of hepatic fibrosis: a pivotal or an incidental player? *Dig Dis Sci* 2006;51:1685-1696.

[20] Lafontan M, Girard J. Impact of visceral adipose tissue on liver metabolism. Part I: heterogeneity of adipose tissue and functional properties of visceral adipose tissue. *Diabetes Metab* 2008;34:317-327.

[21] Marra F, Aleffi S, Bertolani C, Petrai I, Vizzutti F. Adipokines and liver fibrosis. *Eur Rev Med Pharmacol Sci* 2005;9:279-284.

[22] Hu E, Liang P, Spiegelman BM. AdipoQ is a novel adipose-specific gene dysregulated in obesity. *J Biol Chem* 1996;271:10697-10703.

[23] Maeda K, Okubo K, Shimomura I, Funahashi T, Matsuzawa Y, Matsubara K. cDNA cloning and expression of a novel adipose specific collagen-like factor, apM1 (AdiPose Most abundant Gene transcript 1). *Biochem Biophys Res Commun* 1996;221:286-289.

[24] Nakano Y, Tobe T, Choi-Miura NH, Mazda T, Tomita M. Isolation and characterization of GBP28, a novel gelatin-binding protein purified from human plasma. *J Biochem* 1996;120:803-812.

[25] Scherer PE, Williams S, Fogliano M, Baldini G, Lodish HF. A novel serum protein similar to C1q, produced exclusively in adipocytes. *J Biol Chem* 1995;270:26746-26749.

[26] Hu E, Liang P, Spiegelman BM. AdipoQ is a novel adipose-specific gene dysregulated in obesity. *J Biol Chem* 1996;271:10697-10703.

[27] Halberg N, Schraw TD, Wang ZV, Kim JY, Yi J, Hamilton MP, Luby-Phelps K, et al. Systemic fate of the adipocyte-derived factor adiponectin. *Diabetes* 2009;58:1961-1970.

[28] Kadowaki T, Yamauchi T, Kubota N, Hara K, Ueki K, Tobe K. Adiponectin and adiponectin receptors in insulin resistance, diabetes, and the metabolic syndrome. *J Clin Invest* 2006;116:1784-1792.

[29] Arita Y, Kihara S, Ouchi N, Takahashi M, Maeda K, Miyagawa J, Hotta K, et al. Paradoxical decrease of an adipose-specific protein, adiponectin, in obesity. *Biochem Biophys Res Commun* 1999;257:79-83.

[30] Xu A, Chan KW, Hoo RL, Wang Y, Tan KC, Zhang J, Chen B, et al. Testosterone selectively reduces the high molecular weight form of adiponectin by inhibiting its secretion from adipocytes. *J Biol Chem* 2005;280:18073-18080.

[31] Schraw T, Wang ZV, Halberg N, Hawkins M, Scherer PE. Plasma adiponectin complexes have distinct biochemical characteristics. *Endocrinology* 2008;149:2270-2282.

[32] Hotta K, Funahashi T, Bodkin NL, Ortmeyer HK, Arita Y, Hansen BC, Matsuzawa Y. Circulating concentrations of the adipocyte protein adiponectin are decreased in parallel with reduced insulin sensitivity during the progression to type 2 diabetes in rhesus monkeys. *Diabetes* 2001;50:1126-1133.

[33] Adamczak M, Wiecek A, Funahashi T, Chudek J, Kokot F, Matsuzawa Y. Decreased plasma adiponectin concentration in patients with essential hypertension. *Am J Hypertens* 2003;16:72-75.

[34] Lim HS, Tayebjee MH, Tan KT, Patel JV, Macfadyen RJ, Lip GY. Serum adiponectin in coronary heart disease: ethnic differences and relation to coronary artery disease severity. *Heart* 2005;91:1605-1606.

[35] Diez JJ, Iglesias P. The role of the novel adipocyte-derived protein adiponectin in human disease: an update. *Mini Rev Med Chem* 2010;10:856-869.

[36] Choi KM, Ryu OH, Lee KW, Kim HY, Seo JA, Kim SG, Kim NH, et al. Serum adiponectin, interleukin-10 levels and inflammatory markers in the metabolic syndrome. *Diabetes Res Clin Pract* 2007;75:235-240.

[37] Mente A, Razak F, Blankenberg S, Vuksan V, Davis AD, Miller R, Teo K, et al. Ethnic variation in adiponectin and leptin levels and their association with adiposity and insulin resistance. *Diabetes Care;3*3:1629-1634.

[38] Matsuzawa Y, Funahashi T, Kihara S, Shimomura I. Adiponectin and metabolic syndrome. *Arterioscler Thromb Vasc Biol* 2004;24:29-33.

[39] Berg AH, Combs TP, Du X, Brownlee M, Scherer PE. The adipocyte-secreted protein Acrp30 enhances hepatic insulin action. *Nat Med* 2001;7:947-953.

[40] Fruebis J, Tsao TS, Javorschi S, Ebbets-Reed D, Erickson MR, Yen FT, Bihain BE, et al. Proteolytic cleavage product of 30-kDa adipocyte complement-related protein increases fatty acid oxidation in muscle and causes weight loss in mice. *Proc Natl Acad Sci U S A* 2001;98:2005-2010.

[41] Yamauchi T, Kamon J, Waki H, Terauchi Y, Kubota N, Hara K, Mori Y, et al. The fat-derived hormone adiponectin reverses insulin resistance associated with both lipoatrophy and obesity. Nat Med 2001;7:941-946.

[42] Masaki T, Chiba S, Yasuda T, Tsubone T, Kakuma T, Shimomura I, Funahashi T, et al. Peripheral, but not central, administration of adiponectin reduces visceral adiposity and upregulates the expression of uncoupling protein in agouti yellow (Ay/a) obese mice. *Diabetes* 2003;52:2266-2273.

[43] Shibata R, Sato K, Pimentel DR, Takemura Y, Kihara S, Ohashi K, Funahashi T, et al. Adiponectin protects against myocardial ischemia-reperfusion injury through AMPK- and COX-2-dependent mechanisms. *Nat Med* 2005;11:1096-1103. Epub 2005 Sep 1011.

[44] Okamoto Y, Kihara S, Ouchi N, Nishida M, Arita Y, Kumada M, Ohashi K, et al. Adiponectin reduces atherosclerosis in apolipoprotein E-deficient mice. *Circulation* 2002;106:2767-2770.

[45] Xu A, Wang Y, Keshaw H, Xu LY, Lam KS, Cooper GJ. The fat-derived hormone adiponectin alleviates alcoholic and nonalcoholic fatty liver diseases in mice. *J Clin Invest* 2003;112:91-100.

[46] Ohashi K, Kihara S, Ouchi N, Kumada M, Fujita K, Hiuge A, Hibuse T, et al. Adiponectin replenishment ameliorates obesity-related hypertension. *Hypertension* 2006;47:1108-1116.

[47] Shore SA, Terry RD, Flynt L, Xu A, Hug C. Adiponectin attenuates allergen-induced airway inflammation and hyperresponsiveness in mice. *J Allergy Clin Immunol* 2006;118:389-395.

[48] Wang Y, Lam JB, Lam KS, Liu J, Lam MC, Hoo RL, Wu D, et al. Adiponectin modulates the glycogen synthase kinase-3beta/beta-catenin signaling pathway and attenuates mammary tumorigenesis of MDA-MB-231 cells in nude mice. *Cancer Res* 2006;66:11462-11470.

[49] Wang Y, Lam KS, Xu A. Adiponectin as a negative regulator in obesity-related mammary carcinogenesis. *Cell Res* 2007;17:280-282.
[50] Berg AH, Combs TP, Du X, Brownlee M, Scherer PE. The adipocyte-secreted protein Acrp30 enhances hepatic insulin action. *Nat Med* 2001;7:947-953.
[51] Wang Y, Xu A, Knight C, Xu LY, Cooper GJ. Hydroxylation and glycosylation of the four conserved lysine residues in the collagenous domain of adiponectin. Potential role in the modulation of its insulin-sensitizing activity. *J Biol Chem* 2002;277:19521-19529.
[52] Xu A, Wang Y, Keshaw H, Xu LY, Lam KS, Cooper GJ. The fat-derived hormone adiponectin alleviates alcoholic and nonalcoholic fatty liver diseases in mice. *J Clin Invest* 2003;112:91-100.
[53] Xu A, Yin S, Wong L, Chan KW, Lam KS. Adiponectin ameliorates dyslipidemia induced by the human immunodeficiency virus protease inhibitor ritonavir in mice. *Endocrinology* 2004;145:487-494.
[54] Goldstein BJ, Scalia R. Adipokines and vascular disease in diabetes. *Curr Diab Rep* 2007;7:25-33.
[55] Lam KS, Xu A. Adiponectin: protection of the endothelium. *Curr Diab Rep* 2005;5:254-259.
[56] Paz-Filho G, Lim EL, Wong ML, Licinio J. Associations between adipokines and obesity-related cancer. *Front Biosci* 2011;16:1634-1650.
[57] Wang Y, Lam KS, Xu JY, Lu G, Xu LY, Cooper GJ, Xu A. Adiponectin inhibits cell proliferation by interacting with several growth factors in an oligomerization-dependent manner. *J Biol Chem* 2005;280:18341-18347.
[58] Barb D, Pazaitou-Panayiotou K, Mantzoros CS. Adiponectin: a link between obesity and cancer. *Expert Opin Investig Drugs* 2006;15:917-931.
[59] Hug C, Wang J, Ahmad NS, Bogan JS, Tsao TS, Lodish HF. T-cadherin is a receptor for hexameric and high-molecular-weight forms of Acrp30/adiponectin. *Proc Natl Acad Sci* U S A 2004;101:10308-10313.
[60] Yamauchi T, Kamon J, Ito Y, Tsuchida A, Yokomizo T, Kita S, Sugiyama T, et al. Cloning of adiponectin receptors that mediate antidiabetic metabolic effects. *Nature* 2003;423:762-769.
[61] Wang Y, Xu LY, Lam KS, Lu G, Cooper GJ, Xu A. Proteomic characterization of human serum proteins associated with the fat-derived hormone adiponectin. *Proteomics* 2006;6:3862-3870.
[62] Masaie H, Oritani K, Yokota T, Takahashi I, Shirogane T, Ujiie H, Ichii M, et al. Adiponectin binds to chemokines via the globular head and modulates interactions between chemokines and heparan sulfates. *Exp Hematol* 2007;35:947-956.
[63] Wang Y, Lam KS, Yau MH, Xu A. Post-translational modifications of adiponectin: mechanisms and functional implications. *Biochem J* 2008;409:623-633.
[64] Shapiro L, Scherer PE. The crystal structure of a complement-1q family protein suggests an evolutionary link to tumor necrosis factor. *Curr Biol* 1998;8:335-338.
[65] Waki H, Yamauchi T, Kamon J, Ito Y, Uchida S, Kita S, Hara K, et al. Impaired multimerization of human adiponectin mutants associated with diabetes. Molecular structure and multimer formation of adiponectin. *J Biol Chem* 2003;278:40352-40363. Epub 42003 Jul 40323.
[66] Xu A, Chan KW, Hoo RL, Wang Y, Tan KC, Zhang J, Chen B, et al. Testosterone selectively reduces the high molecular weight form of adiponectin by inhibiting its

secretion from adipocytes. *J Biol Chem* 2005;280:18073-18080. Epub 12005 Mar 18079.

[67] Tsao TS, Tomas E, Murrey HE, Hug C, Lee DH, Ruderman NB, Heuser JE, et al. Role of disulfide bonds in Acrp30/Adiponectin structure and signaling specificity: Different oligomers activate different signal transduction pathways. *J Biol Chem* 2003;278:50810-50817.

[68] Pajvani UB, Du X, Combs TP, Berg AH, Rajala MW, Schulthess T, Engel J, et al. Structure-function studies of the adipocyte-secreted hormone Acrp30/adiponectin. Implications fpr metabolic regulation and bioactivity. *J Biol Chem* 2003;278:9073-9085.

[69] Radjainia M, Wang Y, Mitra AK. Structural polymorphism of oligomeric adiponectin visualized by electron microscopy. *J Mol Biol* 2008;381:419-430.

[70] Suzuki S, Wilson-Kubalek EM, Wert D, Tsao TS, Lee DH. The oligomeric structure of high molecular weight adiponectin. *FEBS Lett* 2007;581:809-814.

[71] Pajvani UB, Du X, Combs TP, Berg AH, Rajala MW, Schulthess T, Engel J, et al. Structure-function studies of the adipocyte-secreted hormone Acrp30/adiponectin. Implications fpr metabolic regulation and bioactivity. *J Biol Chem* 2003;278:9073-9085.

[72] Wang Y, Lam KS, Chan L, Chan KW, Lam JB, Lam MC, Hoo RC, et al. Post-translational modifications of the four conserved lysine residues within the collagenous domain of adiponectin are required for the formation of its high molecular weight oligomeric complex. *J Biol Chem* 2006;281:16391-16400.

[73] Kobayashi H, Ouchi N, Kihara S, Walsh K, Kumada M, Abe Y, Funahashi T, et al. Selective suppression of endothelial cell apoptosis by the high molecular weight form of adiponectin. *Circ Res* 2004;94:e27-31. Epub 2004 Jan 2029.

[74] Trujillo ME, Scherer PE. Adiponectin--journey from an adipocyte secretory protein to biomarker of the metabolic syndrome. *J Intern Med* 2005;257:167-175.

[75] Pajvani UB, Hawkins M, Combs TP, Rajala MW, Doebber T, Berger JP, Wagner JA, et al. Complex distribution, not absolute amount of adiponectin, correlates with thiazolidinedione-mediated improvement in insulin sensitivity. *J Biol Chem* 2004;279:12152-12162.

[76] Bobbert T, Rochlitz H, Wegewitz U, Akpulat S, Mai K, Weickert MO, Mohlig M, et al. Changes of adiponectin oligomer composition by moderate weight reduction. *Diabetes* 2005;54:2712-2719.

[77] Salani B, Briatore L, Andraghetti G, Adami GF, Maggi D, Cordera R. High-molecular weight adiponectin isoforms increase after biliopancreatic diversion in obese subjects. *Obesity* (Silver Spring) 2006;14:1511-1514.

[78] Swarbrick MM, Austrheim-Smith IT, Stanhope KL, Van Loan MD, Ali MR, Wolfe BM, Havel PJ. Circulating concentrations of high-molecular-weight adiponectin are increased following Roux-en-Y gastric bypass surgery. *Diabetologia* 2006;49:2552-2558.

[79] Waki H, Yamauchi T, Kamon J, Ito Y, Uchida S, Kita S, Hara K, et al. Impaired multimerization of human adiponectin mutants associated with diabetes. Molecular structure and multimer formation of adiponectin. *J Biol Chem* 2003;278:40352-40363.

[80] Richards AA, Stephens T, Charlton HK, Jones A, Macdonald GA, Prins JB, Whitehead JP. Adiponectin multimerization is dependent on conserved lysines in the collagenous domain: evidence for regulation of multimerization by alterations in posttranslational modifications. *Mol Endocrinol* 2006;20:1673-1687.

[81] Wang Y, Lu G, Wong WP, Vliegenthart JF, Gerwig GJ, Lam KS, Cooper GJ, et al. Proteomic and functional characterization of endogenous adiponectin purified from fetal bovine serum. *Proteomics* 2004;4:3933-3942.

[82] Wang Y, Xu A, Knight C, Xu LY, Cooper GJ. Hydroxylation and glycosylation of the four conserved lysine residues in the collagenous domain of adiponectin. Potential role in the modulation of its insulin-sensitizing activity. *J Biol Chem* 2002;277:19521-19529.

[83] Hug C, Wang J, Ahmad NS, Bogan JS, Tsao TS, Lodish HF. T-cadherin is a receptor for hexameric and high-molecular-weight forms of Acrp30/adiponectin. *Proc Natl Acad Sci* U S A 2004;101:10308-10313. Epub 12004 Jun 10321.

[84] Passoja K, Rautavuoma K, Ala-Kokko L, Kosonen T, Kivirikko KI. Cloning and characterization of a third human lysyl hydroxylase isoform. *Proc Natl Acad Sci* U S A 1998;95:10482-10486.

[85] Wang C, Luosujarvi H, Heikkinen J, Risteli M, Uitto L, Myllyla R. The third activity for lysyl hydroxylase 3: galactosylation of hydroxylysyl residues in collagens in vitro. *Matrix Biol* 2002;21:559-566.

[86] Frizzell N, Rajesh M, Jepson MJ, Nagai R, Carson JA, Thorpe SR, Baynes JW. Succination of thiol groups in adipose tissue proteins in diabetes: succination inhibits polymerization and secretion of adiponectin. *J Biol Chem* 2009;284:25772-25781.

[87] Sato C, Yasukawa Z, Honda N, Matsuda T, Kitajima K. Identification and adipocyte differentiation-dependent expression of the unique disialic acid residue in an adipose tissue-specific glycoprotein, adipo Q. *J Biol Chem* 2001;276:28849-28856. Epub 22001 May 28829.

[88] Richards AA, Colgrave ML, Zhang J, Webster J, Simpson F, Preston E, Wilks D, et al. Sialic acid modification of adiponectin is not required for multimerization or secretion but determines half-life in circulation. *Mol Endocrinol;*24:229-239.

[89] Nobili V, Carter-Kent C, Feldstein AE. The role of lifestyle changes in the management of chronic liver disease. *BMC Med* 2011;9:70.

[90] Williams R. Global challenges in liver disease. *Hepatology* 2006;44:521-526.

[91] Bertolani C, Marra F. The role of adipokines in liver fibrosis. *Pathophysiology* 2008;15:91-101.

[92] Bluher M. Adipose tissue dysfunction in obesity. *Exp Clin Endocrinol Diabetes* 2009;117:241-250.

[93] Tilg H, Hotamisligil GS. Nonalcoholic fatty liver disease: Cytokine-adipokine interplay and regulation of insulin resistance. *Gastroenterology* 2006;131:934-945.

[94] Javor ED, Ghany MG, Cochran EK, Oral EA, DePaoli AM, Premkumar A, Kleiner DE, et al. Leptin reverses nonalcoholic steatohepatitis in patients with severe lipodystrophy. *Hepatology* 2005;41:753-760.

[95] Javor ED, Cochran EK, Musso C, Young JR, Depaoli AM, Gorden P. Long-term efficacy of leptin replacement in patients with generalized lipodystrophy. *Diabetes* 2005;54:1994-2002.

[96] Ding X, Saxena NK, Lin S, Xu A, Srinivasan S, Anania FA. The roles of leptin and adiponectin: a novel paradigm in adipocytokine regulation of liver fibrosis and stellate cell biology. *Am J Pathol* 2005;166:1655-1669.

[97] Crespo J, Cayon A, Fernandez-Gil P, Hernandez-Guerra M, Mayorga M, Dominguez-Diez A, Fernandez-Escalante JC, et al. Gene expression of tumor necrosis factor alpha

and TNF-receptors, p55 and p75, in nonalcoholic steatohepatitis patients. *Hepatology* 2001;34:1158-1163.

[98] Ribeiro PS, Cortez-Pinto H, Sola S, Castro RE, Ramalho RM, Baptista A, Moura MC, et al. Hepatocyte apoptosis, expression of death receptors, and activation of NF-kappaB in the liver of nonalcoholic and alcoholic steatohepatitis patients. *Am J Gastroenterol* 2004;99:1708-1717.

[99] Younossi ZM, Jarrar M, Nugent C, Randhawa M, Afendy M, Stepanova M, Rafiq N, et al. A novel diagnostic biomarker panel for obesity-related nonalcoholic steatohepatitis (NASH). *Obes Surg* 2008;18:1430-1437.

[100] Li Z, Yang S, Lin H, Huang J, Watkins PA, Moser AB, Desimone C, et al. Probiotics and antibodies to TNF inhibit inflammatory activity and improve nonalcoholic fatty liver disease. *Hepatology* 2003;37:343-350.

[101] Deng QG, She H, Cheng JH, French SW, Koop DR, Xiong S, Tsukamoto H. Steatohepatitis induced by intragastric overfeeding in mice. *Hepatology* 2005;42:905-914.

[102] Schnyder-Candrian S, Czarniecki J, Lerondel S, Corpataux J, Ryffel B, Schnyder B. Hepatic steatosis in the absence of tumor necrosis factor in mice. *Cytokine* 2005;32:287-295.

[103] Angulo P. Nonalcoholic fatty liver disease. *N Engl J Med* 2002;346:1221-1231.

[104] Farrell GC. Non-alcoholic steatohepatitis: what is it, and why is it important in the Asia-Pacific region? *J Gastroenterol Hepatol* 2003;18:124-138.

[105] Matteoni CA, Younossi ZM, Gramlich T, Boparai N, Liu YC, McCullough AJ. Nonalcoholic fatty liver disease: a spectrum of clinical and pathological severity. *Gastroenterology* 1999;116:1413-1419.

[106] Jeong SK, Kim YK, Park JW, Shin YJ, Kim DS. Impact of visceral fat on the metabolic syndrome and nonalcoholic fatty liver disease. *J Korean Med Sci* 2008;23:789-795.

[107] van der Poorten D, Milner KL, Hui J, Hodge A, Trenell MI, Kench JG, London R, et al. Visceral fat: a key mediator of steatohepatitis in metabolic liver disease. *Hepatology* 2008;48:449-457.

[108] Farrell GC, Larter CZ. Nonalcoholic fatty liver disease: from steatosis to cirrhosis. *Hepatology* 2006;43:S99-S112.

[109] Jakobsen MU, Berentzen T, Sorensen TI, Overvad K. Abdominal obesity and fatty liver. *Epidemiol Rev* 2007;29:77-87.

[110] Ong JP, Pitts A, Younossi ZM. Increased overall mortality and liver-related mortality in non-alcoholic fatty liver disease. *J Hepatol* 2008;49:608-612.

[111] Targher G, Marra F, Marchesini G. Increased risk of cardiovascular disease in non-alcoholic fatty liver disease: causal effect or epiphenomenon? *Diabetologia* 2008;51:1947-1953.

[112] Watanabe S, Yaginuma R, Ikejima K, Miyazaki A. Liver diseases and metabolic syndrome. *J Gastroenterol* 2008;43:509-518.

[113] Machado M, Cortez-Pinto H. Non-alcoholic steatohepatitis and metabolic syndrome. *Curr Opin Clin Nutr Metab Care* 2006;9:637-642.

[114] Shibata M, Kihara Y, Taguchi M, Tashiro M, Otsuki M. Nonalcoholic fatty liver disease is a risk factor for type 2 diabetes in middle-aged Japanese men. *Diabetes Care* 2007;30:2940-2944.

[115] Lizardi-Cervera J, Aguilar-Zapata D. Nonalcoholic fatty liver disease and its association with cardiovascular disease. *Ann Hepatol* 2009;8 Suppl 1:S40-43.

[116] Kamada Y, Takehara T, Hayashi N. Adipocytokines and liver disease. *J Gastroenterol* 2008;43:811-822.

[117] Pisto P, Ukkola O, Santaniemi M, Kesaniemi YA. *Plasma adiponectin-an independent indicator of liver fat accumulation.* Metabolism 2011.

[118] Lopez-Bermejo A, Botas P, Funahashi T, Delgado E, Kihara S, Ricart W, Fernandez-Real JM. Adiponectin, hepatocellular dysfunction and insulin sensitivity. *Clin Endocrinol* (Oxf) 2004;60:256-263.

[119] Targher G, Bertolini L, Scala L, Poli F, Zenari L, Falezza G. Decreased plasma adiponectin concentrations are closely associated with nonalcoholic hepatic steatosis in obese individuals. *Clin Endocrinol* (Oxf) 2004;61:700-703.

[120] Vuppalanchi R, Marri S, Kolwankar D, Considine RV, Chalasani N. Is adiponectin involved in the pathogenesis of nonalcoholic steatohepatitis? A preliminary human study. *J Clin Gastroenterol* 2005;39:237-242.

[121] Bugianesi E, Pagotto U, Manini R, Vanni E, Gastaldelli A, de Iasio R, Gentilcore E, et al. Plasma adiponectin in nonalcoholic fatty liver is related to hepatic insulin resistance and hepatic fat content, not to liver disease severity. *J Clin Endocrinol Metab* 2005;90:3498-3504.

[122] Musso G, Gambino R, Durazzo M, Biroli G, Carello M, Faga E, Pacini G, et al. Adipokines in NASH: postprandial lipid metabolism as a link between adiponectin and liver disease. *Hepatology* 2005;42:1175-1183.

[123] Hui JM, Hodge A, Farrell GC, Kench JG, Kriketos A, George J. Beyond insulin resistance in NASH: TNF-alpha or adiponectin? *Hepatology* 2004;40:46-54.

[124] Shimada M, Kawahara H, Ozaki K, Fukura M, Yano H, Tsuchishima M, Tsutsumi M, et al. Usefulness of a combined evaluation of the serum adiponectin level, HOMA-IR, and serum type IV collagen 7S level to predict the early stage of nonalcoholic steatohepatitis. *Am J Gastroenterol* 2007;102:1931-1938.

[125] Bajaj M, Suraamornkul S, Piper P, Hardies LJ, Glass L, Cersosimo E, Pratipanawatr T, et al. Decreased plasma adiponectin concentrations are closely related to hepatic fat content and hepatic insulin resistance in pioglitazone-treated type 2 diabetic patients. *J Clin Endocrinol Metab* 2004;89:200-206.

[126] Aguilar-Salinas CA, Garcia EG, Robles L, Riano D, Ruiz-Gomez DG, Garcia-Ulloa AC, Melgarejo MA, et al. High adiponectin concentrations are associated with the metabolically healthy obese phenotype. *J Clin Endocrinol Metab* 2008;93:4075-4079.

[127] Wang ZL, Xia B, Shrestha U, Jiang L, Ma CW, Chen Q, Chen H, et al. Correlation between adiponectin polymorphisms and non-alcoholic fatty liver disease with or without metabolic syndrome in Chinese population. *J Endocrinol Invest* 2008;31:1086-1091.

[128] Musso G, Gambino R, De Michieli F, Durazzo M, Pagano G, Cassader M. Adiponectin gene polymorphisms modulate acute adiponectin response to dietary fat: Possible pathogenetic role in NASH. *Hepatology* 2008;47:1167-1177.

[129] Kotronen A, Yki-Jarvinen H, Aminoff A, Bergholm R, Pietilainen KH, Westerbacka J, Talmud PJ, et al. Genetic variation in the ADIPOR2 gene is associated with liver fat content and its surrogate markers in three independent cohorts. *Eur J Endocrinol* 2009;160:593-602.

[130] Stefan N, Machicao F, Staiger H, Machann J, Schick F, Tschritter O, Spieth C, et al. Polymorphisms in the gene encoding adiponectin receptor 1 are associated with insulin resistance and high liver fat. *Diabetologia* 2005;48:2282-2291.

[131] Anty R, Lemoine M. Liver fibrogenesis and metabolic factors. *Clin Res Hepatol Gastroenterol* 2011;35 Suppl 1:S10-20.

[132] Chiang DJ, Pritchard MT, Nagy LE. Obesity, diabetes mellitus, and liver fibrosis. Am J Physiol Gastrointest Liver *Physiol* 2011;300:G697-702.

[133] Sorrentino P, Terracciano L, D'Angelo S, Ferbo U, Bracigliano A, Vecchione R. Predicting fibrosis worsening in obese patients with NASH through parenchymal fibronectin, HOMA-IR, and hypertension. A*m J Gastroenterol* 2009;105:336-344.

[134] Lanthier N, Horsmans Y, Leclercq IA. The metabolic syndrome: how it may influence hepatic stellate cell activation and hepatic fibrosis. *Curr Opin Clin Nutr Metab Care* 2009;12:404-411.

[135] Nazal L, Riquelme A, Solis N, Pizarro M, Escalona A, Burotto M, Mendez JI, et al. Hypoadiponectinemia and its association with liver fibrosis in morbidly obese patients. *Obes Surg* 2010;20:1400-1407.

[136] Savvidou S, Hytiroglou P, Orfanou-Koumerkeridou H, Panderis A, Frantzoulis P, Goulis J. Low serum adiponectin levels are predictive of advanced hepatic fibrosis in patients with NAFLD. *J Clin Gastroenterol* 2009;43:765-772.

[137] Tokushige K, Hashimoto E, Noto H, Yatsuji S, Taniai M, Torii N, Shiratori K. Influence of adiponectin gene polymorphisms in Japanese patients with non-alcoholic fatty liver disease. *J Gastroenterol* 2009;44:976-982.

[138] Namvaran F, Rahimi-Moghaddam P, Azarpira N, Nikeghbalian S. The association between adiponectin (+45T/G) and adiponectin receptor-2 (+795G/A) single nucleotide polymorphisms with cirrhosis in Iranian population. *Mol Biol Rep* 2011.

[139] Kaser S, Moschen A, Cayon A, Kaser A, Crespo J, Pons-Romero F, Ebenbichler CF, et al. Adiponectin and its receptors in non-alcoholic steatohepatitis. *Gut* 2005;54:117-121.

[140] Ma H, Gomez V, Lu L, Yang X, Wu X, Xiao SY. Expression of adiponectin and its receptors in livers of morbidly obese patients with non-alcoholic fatty liver disease. *J Gastroenterol Hepatol* 2009;24:233-237.

[141] Asada K, Yoshiji H, Noguchi R, Ikenaka Y, Kitade M, Kaji K, Yoshii J, et al. Crosstalk between high-molecular-weight adiponectin and T-cadherin during liver fibrosis development in rats. *Int J Mol Med* 2007;20:725-729.

[142] Tacke F, Wustefeld T, Horn R, Luedde T, Srinivas Rao A, Manns MP, Trautwein C, et al. High adiponectin in chronic liver disease and cholestasis suggests biliary route of adiponectin excretion in vivo. *J Hepatol* 2005;42:666-673.

[143] Friedman SL. Liver fibrosis -- from bench to bedside. *J Hepatol* 2003;38 Suppl 1:S38-53.

[144] Salman TA, Allam N, Azab GI, Shaarawy AA, Hassouna MM, El-Haddad OM. Study of adiponectin in chronic liver disease and cholestasis. *Hepatol Int* 2011;4:767-774.

[145] Record CO. Glucose and insulin metabolism in cirrhosis. *Adv Exp Med Biol* 1997;420:229-233.

[146] Tietge UJ, Boker KH, Manns MP, Bahr MJ. Elevated circulating adiponectin levels in liver cirrhosis are associated with reduced liver function and altered hepatic hemodynamics. *Am J Physiol Endocrinol Metab* 2004;287:E82-89.

[147] Tilg H, Wilmer A, Vogel W, Herold M, Nolchen B, Judmaier G, Huber C. Serum levels of cytokines in chronic liver diseases. *Gastroenterology* 1992;103:264-274.

[148] Kaser S, Moschen A, Kaser A, Ludwiczek O, Ebenbichler CF, Vogel W, Jaschke W, et al. Circulating adiponectin reflects severity of liver disease but not insulin sensitivity in liver cirrhosis. *J Intern Med* 2005;258:274-280.

[149] Lo Iacono O, Venezia G, Petta S, Mineo C, De Lisi S, Di Marco V, Rodolico V, et al. The impact of insulin resistance, serum adipocytokines and visceral obesity on steatosis and fibrosis in patients with chronic hepatitis C. *Aliment Pharmacol Ther* 2007;25:1181-1191.

[150] Fartoux L, Poujol-Robert A, Guechot J, Wendum D, Poupon R, Serfaty L. Insulin resistance is a cause of steatosis and fibrosis progression in chronic hepatitis C. *Gut* 2005;54:1003-1008.

[151] Ashour E, Samy N, Sayed M, Imam A. The relationship between serum adiponectin and steatosis in patients with chronic hepatitis C genotype-4. *Clin Lab* 2010;56:103-110.

[152] Hung CH, Lee CM, Chen CH, Hu TH, Jiang SR, Wang JH, Lu SN, et al. Association of inflammatory and anti-inflammatory cytokines with insulin resistance in chronic hepatitis C. *Liver Int* 2009;29:1086-1093.

[153] Jonsson JR, Moschen AR, Hickman IJ, Richardson MM, Kaser S, Clouston AD, Powell EE, et al. Adiponectin and its receptors in patients with chronic hepatitis C. *J Hepatol* 2005;43:929-936.

[154] Patel K, Thompson AJ, Chuang WL, Lee CM, Peng CY, Shanmuganathan G, Thongsawat S, et al. Insulin resistance is independently associated with significant hepatic fibrosis in Asian chronic hepatitis C genotype 2 or 3 patients. *J Gastroenterol Hepatol* 2011;26:1182-1188.

[155] Liu CJ, Chen PJ, Jeng YM, Huang WL, Yang WS, Lai MY, Kao JH, et al. Serum adiponectin correlates with viral characteristics but not histologic features in patients with chronic hepatitis C. *J Hepatol* 2005;43:235-242.

[156] Wang AY, Hickman IJ, Richards AA, Whitehead JP, Prins JB, Macdonald GA. High molecular weight adiponectin correlates with insulin sensitivity in patients with hepatitis C genotype 3, but not genotype 1 infection. *Am J Gastroenterol* 2005;100:2717-2723.

[157] Zografos TA, Liaskos C, Rigopoulou EI, Togousidis E, Makaritsis K, Germenis A, Dalekos GN. Adiponectin: a new independent predictor of liver steatosis and response to IFN-alpha treatment in chronic hepatitis C. *Am J Gastroenterol* 2008;103:605-614.

[158] Hsu CS, Liu CJ, Liu CH, Wang CC, Chen CL, Lai MY, Chen PJ, et al. High hepatitis C viral load is associated with insulin resistance in patients with chronic hepatitis C. *Liver Int* 2008;28:271-277.

[159] Corbetta S, Redaelli A, Pozzi M, Bovo G, Ratti L, Redaelli E, Pellegrini C, et al. Fibrosis is associated with adiponectin resistance in chronic hepatitis C virus infection. *Eur J Clin Invest* 2011;41:898-905.

[160] Latif HA, Assal HS, Mahmoud M, Rasheed WI. Role of serum adiponectin level in the development of liver cirrhosis in patients with hepatitis C virus. *Clin Exp Med* 2010;11:123-129.

[161] Kukla M, Mazur W, Buldak RJ, Zwirska-Korczala K. Potential role of leptin, adiponectin and the novel adipokines-visfatin, chemerin and vaspin-in chronic hepatitis. *Mol Med* 2011.

[162] Arrese M, Riquelme A, Soza A. Insulin resistance, hepatic steatosis and hepatitis C: a complex relationship with relevant clinical implications. *Ann Hepatol* 2010;9 Suppl:112-118.

[163] Wong VW, Wong GL, Yu J, Choi PC, Chan AW, Chan HY, Chu ES, et al. Interaction of adipokines and hepatitis B virus on histological liver injury in the Chinese. *Am J Gastroenterol* 2010;105:132-138.

[164] Liu CJ, Chen PJ, Lai MY, Liu CH, Chen CL, Kao JH, Chen DS. High serum adiponectin correlates with advanced liver disease in patients with chronic hepatitis B virus infection. *Hepatol Int* 2009;3:364-370.

[165] Hui CK, Zhang HY, Lee NP, Chan W, Yueng YH, Leung KW, Lu L, et al. Serum adiponectin is increased in advancing liver fibrosis and declines with reduction in fibrosis in chronic hepatitis B. *J Hepatol* 2007;47:191-202.

[166] Tilg H. Adipocytokines in nonalcoholic fatty liver disease: key players regulating steatosis, inflammation and fibrosis. *Curr Pharm Des* 2010;16:1893-1895.

[167] Veldt BJ, Chen W, Heathcote EJ, Wedemeyer H, Reichen J, Hofmann WP, de Knegt RJ, et al. Increased risk of hepatocellular carcinoma among patients with hepatitis C cirrhosis and diabetes mellitus. *Hepatology* 2008;47:1856-1862.

[168] Calle EE, Rodriguez C, Walker-Thurmond K, Thun MJ. Overweight, obesity, and mortality from cancer in a prospectively studied cohort of U.S. adults. *N Engl J Med* 2003;348:1625-1638.

[169] Hassan MM, Curley SA, Li D, Kaseb A, Davila M, Abdalla EK, Javle M, et al. Association of diabetes duration and diabetes treatment with the risk of hepatocellular carcinoma. *Cancer* 2010;116:1938-1946.

[170] Nkontchou G, Bastard JP, Ziol M, Aout M, Cosson E, Ganne-Carrie N, Grando-Lemaire V, et al. Insulin resistance, serum leptin, and adiponectin levels and outcomes of viral hepatitis C cirrhosis. *J Hepatol* 2010;53:827-833.

[171] Hung CH, Wang JH, Hu TH, Chen CH, Chang KC, Yen YH, Kuo YH, et al. Insulin resistance is associated with hepatocellular carcinoma in chronic hepatitis C infection. *World J Gastroenterol* 2010;16:2265-2271.

[172] Man K, Ng KT, Xu A, Cheng Q, Lo CM, Xiao JW, Sun BS, et al. Suppression of liver tumor growth and metastasis by adiponectin in nude mice through inhibition of tumor angiogenesis and downregulation of Rho kinase/IFN-inducible protein 10/matrix metalloproteinase 9 signaling. *Clin Cancer Res* 2010;16:967-977.

[173] Saxena NK, Fu PP, Nagalingam A, Wang J, Handy J, Cohen C, Tighiouart M, et al. Adiponectin modulates C-jun N-terminal kinase and mammalian target of rapamycin and inhibits hepatocellular carcinoma. *Gastroenterology* 2010;139:1762-1773, 1773 e1761-1765.

[174] Arano T, Nakagawa H, Tateishi R, Ikeda H, Uchino K, Enooku K, Goto E, et al. Serum level of adiponectin and the risk of liver cancer development in chronic hepatitis C patients. *Int J Cancer* 2010.

[175] Fukushima N, Kuromatsu R, Arinaga-Hino T, Ando E, Takata A, Sumie S, Nakano M, et al. Adipocytokine involvement in hepatocellular carcinoma after sustained response to interferon for chronic hepatitis C. *Hepatol Res* 2010;40:911-922.

[176] Adachi Y, Takeuchi T, Sonobe H, Ohtsuki Y. An adiponectin receptor, T-cadherin, was selectively expressed in intratumoral capillary endothelial cells in hepatocellular carcinoma: possible cross talk between T-cadherin and FGF-2 pathways. *Virchows Arch* 2006;448:311-318.

[177] Wang Y, Zhou M, Lam KS, Xu A. Protective roles of adiponectin in obesity-related fatty liver diseases: mechanisms and therapeutic implications. *Arq Bras Endocrinol Metabol* 2009;53:201-212.

[178] Ikejima K, Okumura K, Kon K, Takei Y, Sato N. Role of adipocytokines in hepatic fibrogenesis. *J Gastroenterol Hepatol* 2007;22 Suppl 1:S87-92.

[179] Kamada Y, Tamura S, Kiso S, Matsumoto H, Saji Y, Yoshida Y, Fukui K, et al. Enhanced carbon tetrachloride-induced liver fibrosis in mice lacking adiponectin. *Gastroenterology* 2003;125:1796-1807.

[180] Masaki T, Chiba S, Tatsukawa H, Yasuda T, Noguchi H, Seike M, Yoshimatsu H. Adiponectin protects LPS-induced liver injury through modulation of TNF-alpha in KK-Ay obese mice. *Hepatology* 2004;40:177-184.

[181] Matsumoto H, Tamura S, Kamada Y, Kiso S, Fukushima J, Wada A, Maeda N, et al. Adiponectin deficiency exacerbates lipopolysaccharide/D-galactosamine-induced liver injury in mice. *World J Gastroenterol* 2006;12:3352-3358.

[182] Thakur V, Pritchard MT, McMullen MR, Nagy LE. Adiponectin normalizes LPS-stimulated TNF-alpha production by rat Kupffer cells after chronic ethanol feeding. *Am J Physiol Gastrointest Liver Physiol* 2006;290:G998-1007.

[183] Larter CZ, Yeh MM, Williams J, Bell-Anderson KS, Farrell GC. MCD-induced steatohepatitis is associated with hepatic adiponectin resistance and adipogenic transformation of hepatocytes. *J Hepatol* 2008;49:407-416.

[1] Zhou M, Xu A, Tam PK, Lam KS, Chan L, Hoo RL, Liu J, et al. Mitochondrial dysfunction contributes to the increased vulnerabilities of adiponectin knockout mice to liver injury. *Hepatology* 2008;48:1087-1096.

[184] Asano T, Watanabe K, Kubota N, Gunji T, Omata M, Kadowaki T, Ohnishi S. Adiponectin knockout mice on high fat diet develop fibrosing steatohepatitis. *J Gastroenterol Hepatol* 2009;24:1669-1676.

[185] Zhou M, Xu A, Lam KS, Tam PK, Che CM, Chan L, Lee IK, et al. Rosiglitazone promotes fatty acyl CoA accumulation and excessive glycogen storage in livers of mice without adiponectin. *J Hepatol* 2010.

[186] Kamada Y, Matsumoto H, Tamura S, Fukushima J, Kiso S, Fukui K, Igura T, et al. Hypoadiponectinemia accelerates hepatic tumor formation in a nonalcoholic steatohepatitis mouse model. *J Hepatol* 2007;47:556-564.

[187] Nishihara T, Baba M, Matsuda M, Inoue M, Nishizawa Y, Fukuhara A, Araki H, et al. Adiponectin deficiency enhances colorectal carcinogenesis and liver tumor formation induced by azoxymethane in mice. *World J Gastroenterol* 2008;14:6473-6480.

[188] Park PH, Thakur V, Pritchard MT, McMullen MR, Nagy LE. Regulation of Kupffer cell activity during chronic ethanol exposure: role of adiponectin. *J Gastroenterol Hepatol* 2006;21 Suppl 3:S30-33.

[189] Choi S, Diehl AM. Role of inflammation in nonalcoholic steatohepatitis. *Curr Opin Gastroenterol* 2005;21:702-707.

[190] Jarrar MH, Baranova A, Collantes R, Ranard B, Stepanova M, Bennett C, Fang Y, et al. Adipokines and cytokines in non-alcoholic fatty liver disease. *Aliment Pharmacol Ther* 2008;27:412-421.

[191] Ouchi N, Kihara S, Arita Y, Okamoto Y, Maeda K, Kuriyama H, Hotta K, et al. Adiponectin, an adipocyte-derived plasma protein, inhibits endothelial NF-kappaB signaling through a cAMP-dependent pathway. *Circulation* 2000;102:1296-1301.

[192] Kumada M, Kihara S, Ouchi N, Kobayashi H, Okamoto Y, Ohashi K, Maeda K, et al. Adiponectin specifically increased tissue inhibitor of metalloproteinase-1 through interleukin-10 expression in human macrophages. *Circulation* 2004;109:2046-2049.

[193] Wolf AM, Wolf D, Rumpold H, Enrich B, Tilg H. Adiponectin induces the anti-inflammatory cytokines IL-10 and IL-1RA in human leukocytes. *Biochem Biophys Res Commun* 2004;323:630-635.

[194] Wulster-Radcliffe MC, Ajuwon KM, Wang J, Christian JA, Spurlock ME. Adiponectin differentially regulates cytokines in porcine macrophages. *Biochem Biophys Res Commun* 2004;316:924-929.

[195] Yamaguchi N, Argueta JG, Masuhiro Y, Kagishita M, Nonaka K, Saito T, Hanazawa S, et al. Adiponectin inhibits Toll-like receptor family-induced signaling. *FEBS Lett* 2005;579:6821-6826.

[196] Caligiuri A, Bertolani C, Guerra CT, Aleffi S, Galastri S, Trappoliere M, Vizzutti F, et al. Adenosine monophosphate-activated protein kinase modulates the activated phenotype of hepatic stellate cells. *Hepatology* 2008;47:668-676.

[197] Adachi M, Brenner DA. High molecular weight adiponectin inhibits proliferation of hepatic stellate cells via activation of adenosine monophosphate-activated protein kinase. *Hepatology* 2008;47:677-685.

[198] Tomita K, Oike Y, Teratani T, Taguchi T, Noguchi M, Suzuki T, Mizutani A, et al. Hepatic AdipoR2 signaling plays a protective role against progression of nonalcoholic steatohepatitis in mice. *Hepatology* 2008;48:458-473.

[199] Wanninger J, Neumeier M, Weigert J, Bauer S, Weiss TS, Schaffler A, Krempl C, et al. Adiponectin-stimulated CXCL8 release in primary human hepatocytes is regulated by ERK1/ERK2, p38 MAPK, NF-kappaB, and STAT3 signaling pathways. *Am J Physiol Gastrointest Liver Physiol* 2009;297:G611-618.

[200] Wedemeyer I, Bechmann LP, Odenthal M, Jochum C, Marquitan G, Drebber U, Gerken G, et al. Adiponectin inhibits steatotic CD95/Fas up-regulation by hepatocytes: therapeutic implications for hepatitis *C. J Hepatol* 2009;50:140-149.

[201] Schaffler A, Scholmerich J, Buchler C. Mechanisms of disease: adipocytokines and visceral adipose tissue--emerging role in nonalcoholic fatty liver disease. *Nat Clin Pract Gastroenterol Hepatol* 2005;2:273-280.

[202] Cheng KK, Lam KS, Wang Y, Huang Y, Carling D, Wu D, Wong C, et al. Adiponectin-induced endothelial nitric oxide synthase activation and nitric oxide production are mediated by APPL1 in endothelial cells. *Diabetes* 2007;56:1387-1394.

[203] Mao X, Kikani CK, Riojas RA, Langlais P, Wang L, Ramos FJ, Fang Q, et al. APPL1 binds to adiponectin receptors and mediates adiponectin signalling and function. *Nat Cell Biol* 2006;8:516-523.

[204] Woods A, Azzout-Marniche D, Foretz M, Stein SC, Lemarchand P, Ferre P, Foufelle F, et al. Characterization of the role of AMP-activated protein kinase in the regulation of

glucose-activated gene expression using constitutively active and dominant negative forms of the kinase. *Mol Cell Biol* 2000;20:6704-6711.

[205] Yamauchi T, Kamon J, Minokoshi Y, Ito Y, Waki H, Uchida S, Yamashita S, et al. Adiponectin stimulates glucose utilization and fatty-acid oxidation by activating AMP-activated protein kinase. *Nat Med* 2002;8:1288-1295.

[206] Tomita K, Tamiya G, Ando S, Kitamura N, Koizumi H, Kato S, Horie Y, et al. AICAR, an AMPK activator, has protective effects on alcohol-induced fatty liver in rats. *Alcohol Clin Exp Res* 2005;29:240S-245S.

[207] You M, Considine RV, Leone TC, Kelly DP, Crabb DW. Role of adiponectin in the protective action of dietary saturated fat against alcoholic fatty liver in mice. *Hepatology* 2005;42:568-577.

[208] Fruchart JC, Duriez P, Staels B. Peroxisome proliferator-activated receptor-alpha activators regulate genes governing lipoprotein metabolism, vascular inflammation and atherosclerosis. *Curr Opin Lipidol* 1999;10:245-257.

[209] Browning JD, Horton JD. Molecular mediators of hepatic steatosis and liver injury. *J Clin Invest* 2004;114:147-152.

[210] Tomita K, Tamiya G, Ando S, Ohsumi K, Chiyo T, Mizutani A, Kitamura N, et al. Tumour necrosis factor alpha signalling through activation of Kupffer cells plays an essential role in liver fibrosis of non-alcoholic steatohepatitis in mice. *Gut* 2006;55:415-424.

[211] Maeda N, Shimomura I, Kishida K, Nishizawa H, Matsuda M, Nagaretani H, Furuyama N, et al. Diet-induced insulin resistance in mice lacking adiponectin/ACRP30. *Nat Med* 2002;8:731-737.

[212] Kundu TK, Hille R, Velayutham M, Zweier JL. Characterization of superoxide production from aldehyde oxidase: an important source of oxidants in biological tissues. *Arch Biochem Biophys* 2007;460:113-121.

[213] Neumeier M, Weigert J, Schaffler A, Weiss TS, Schmidl C, Buttner R, Bollheimer C, et al. Aldehyde oxidase 1 is highly abundant in hepatic steatosis and is downregulated by adiponectin and fenofibric acid in hepatocytes in vitro. *Biochem Biophys Res Commun* 2006;350:731-735.

[214] Yamauchi T, Nio Y, Maki T, Kobayashi M, Takazawa T, Iwabu M, Okada-Iwabu M, et al. Targeted disruption of AdipoR1 and AdipoR2 causes abrogation of adiponectin binding and metabolic actions. *Nat Med* 2007;13:332-339.

[215] Negre-Salvayre A, Hirtz C, Carrera G, Cazenave R, Troly M, Salvayre R, Penicaud L, et al. A role for uncoupling protein-2 as a regulator of mitochondrial hydrogen peroxide generation. *FASEB J* 1997;11:809-815.

[216] Bai Y, Onuma H, Bai X, Medvedev AV, Misukonis M, Weinberg JB, Cao W, et al. Persistent nuclear factor-kappa B activation in Ucp2-/- mice leads to enhanced nitric oxide and inflammatory cytokine production. *J Biol Chem* 2005;280:19062-19069.

[217] Kobayashi K, Inoguchi T, Sonoda N, Sekiguchi N, Nawata H. Adiponectin inhibits the binding of low-density lipoprotein to biglycan, a vascular proteoglycan. *Biochem Biophys Res Commun* 2005;335:66-70.

[218] Okamoto Y, Arita Y, Nishida M, Muraguchi M, Ouchi N, Takahashi M, Igura T, et al. An adipocyte-derived plasma protein, adiponectin, adheres to injured vascular walls. *Horm Metab Res* 2000;32:47-50.

[219] Takemura Y, Ouchi N, Shibata R, Aprahamian T, Kirber MT, Summer RS, Kihara S, et al. Adiponectin modulates inflammatory reactions via calreticulin receptor-dependent clearance of early apoptotic bodies. *J Clin Invest* 2007;117:375-386.
[220] Sharma D, Wang J, Fu PP, Sharma S, Nagalingam A, Mells J, Handy J, et al. Adiponectin antagonizes the oncogenic actions of leptin in hepatocellular carcinogenesis. *Hepatology* 2010;52:1713-1722.
[221] Neumeier M, Sigruener A, Eggenhofer E, Weigert J, Weiss TS, Schaeffler A, Schlitt HJ, et al. High molecular weight adiponectin reduces apolipoprotein B and E release in human hepatocytes. *Biochem Biophys Res Commun* 2007;352:543-548.
[222] Liu Y, Retnakaran R, Hanley A, Tungtrongchitr R, Shaw C, Sweeney G. Total and high molecular weight but not trimeric or hexameric forms of adiponectin correlate with markers of the metabolic syndrome and liver injury in Thai subjects. *J Clin Endocrinol Metab* 2007;92:4313-4318.
[223] Phillips SA, Kung J, Ciaraldi TP, Choe C, Christiansen L, Mudaliar S, Henry RR. Selective regulation of cellular and secreted multimeric adiponectin by antidiabetic therapies in humans. *Am J Physiol Endocrinol Metab* 2009;297:E767-773.
[224] Wang ZV, Schraw TD, Kim JY, Khan T, Rajala MW, Follenzi A, Scherer PE. Secretion of the adipocyte-specific secretory protein adiponectin critically depends on thiol-mediated protein retention. *Mol Cell Biol* 2007;27:3716-3731.
[225] Alberini CM, Bet P, Milstein C, Sitia R. Secretion of immunoglobulin M assembly intermediates in the presence of reducing agents. *Nature* 1990;347:485-487.
[226] Anelli T, Alessio M, Bachi A, Bergamelli L, Bertoli G, Camerini S, Mezghrani A, et al. Thiol-mediated protein retention in the endoplasmic reticulum: the role of ERp44. *Embo J* 2003;22:5015-5022.
[227] Liu M, Zhou L, Xu A, Lam KS, Wetzel MD, Xiang R, Zhang J, et al. A disulfide-bond A oxidoreductase-like protein (DsbA-L) regulates adiponectin multimerization. *Proc Natl Acad Sci* U S A 2008;105:18302-18307.
[228] Wang ZV, Scherer PE. DsbA-L is a versatile player in adiponectin secretion. *Proc Natl Acad Sci* U S A 2008;105:18077-18078.
[229] Qiang L, Wang H, Farmer SR. Adiponectin secretion is regulated by SIRT1 and the endoplasmic reticulum oxidoreductase Ero1-L alpha. *Mol Cell Biol* 2007;27:4698-4707.
[230] Xie L, Boyle D, Sanford D, Scherer PE, Pessin JE, Mora S. Intracellular trafficking and secretion of adiponectin is dependent on GGA-coated vesicles. *J Biol Chem* 2006;281:7253-7259.
[231] Shah K, Stufflebam A, Hilton TN, Sinacore DR, Klein S, Villareal DT. Diet and exercise interventions reduce intrahepatic fat content and improve insulin sensitivity in obese older adults. *Obesity* (Silver Spring) 2009;17:2162-2168.
[232] Petersen KF, Dufour S, Befroy D, Lehrke M, Hendler RE, Shulman GI. Reversal of nonalcoholic hepatic steatosis, hepatic insulin resistance, and hyperglycemia by moderate weight reduction in patients with type 2 diabetes. *Diabetes* 2005;54:603-608.
[233] Dixon JB, Bhathal PS, Hughes NR, O'Brien PE. Nonalcoholic fatty liver disease: Improvement in liver histological analysis with weight loss. *Hepatology* 2004;39:1647-1654.

[234] Promrat K, Kleiner DE, Niemeier HM, Jackvony E, Kearns M, Wands JR, Fava JL, et al. Randomized controlled trial testing the effects of weight loss on nonalcoholic steatohepatitis. *Hepatology* 2010;51:121-129.

[235] Sofer E, Boaz M, Matas Z, Mashavi M, Shargorodsky M. Treatment with insulin sensitizer metformin improves arterial properties, metabolic parameters, and liver function in patients with nonalcoholic fatty liver disease: a randomized, placebo-controlled trial. *Metabolism* 2011;60:1278-1284.

[236] Matafome P, Louro T, Rodrigues L, Crisostomo J, Nunes E, Amaral C, Monteiro P, et al. Metformin and atorvastatin combination further protect the liver in type 2 diabetes with hyperlipidaemia. *Diabetes Metab Res Rev* 2011;27:54-62.

[237] Yki-Jarvinen H. Thiazolidinediones and the liver in humans. *Curr Opin Lipidol* 2009;20:477-483.

[238] Xu A, Wang H, Hoo RL, Sweeney G, Vanhoutte PM, Wang Y, Wu D, et al. Selective elevation of adiponectin production by the natural compounds derived from a medicinal herb alleviates insulin resistance and glucose intolerance in obese mice. *Endocrinology* 2009;150:625-633.

[239] Xu A, Wang H, Hoo RL, Sweeney G, Vanhoutte PM, Wang Y, Wu D, et al. Selective Elevation of Adiponectin Production by the Natural Compounds Derived from a Medicinal Herb Alleviates Insulin Resistance and Glucose Intolerance in Obese Mice. *Endocrinology* 2008.

In: Adiponectin: Production, Regulation and Roles in Disease ISBN: 978-1-61942-165-3
Editor: Masayoshi Yamaguchi

Chapter VI

SEPSIS AND ADIPONECTIN

Luca Siracusano* and Viviana Girasole
Department of Neuroscience, Psychiatric and Anesthesiological Sciences,
University of Messina, School of Medicine, Policlinico Universitario G. Martino, Italy

ABSTRACT

Adiponectin is an adipokine provided with antiatherosclerotic, insulin sensitizing, anti-inflammatory and cardioprotective actions.

The adipose tissue, considered in the past as a long-term energy storage organ with a relatively reduced metabolic activity, is now considered as an important player in the regulation of the metabolism and of the immune function carrying out a fundamental role in the outcome of the critically ill patient.

The finding of a paradoxical reduction of adiponectin, produced almost exclusively in the adipose tissue, in obesity, diabetes, metabolic syndrome, sepsis and critical diseases have led to study the significance of the serum levels of this adipokine showing they are strictly related to the outcome in critically ill patients.

Considering that sepsis is characterized by a disregulated inflammatory reaction and the important role that the reduced insulin sensitivity has been attributed to in the septic patient it is not surprising that hypoadiponectinemia may have a prognostic role in this setting by influencing metabolism and immune system.

Adiponectin is downregulated during experimental ischemia-reperfusion injury but it is provided with a powerful cardioprotective action through the activation of the enzymes of the Reperfusion Injury Salvage Kinase (RISK) pathway Akt\PI3K and through the inhibition of mitochondrial permeability pore transition . The cardioprotection conferred by adiponectin is also mediated by its ability to induce myocardial preconditioning.

The altered secretion of adipokines in sepsis is probably connected to the dysfunction adipose tissue undergoes in this disease characterized by reduced insulin sensitivity, inflammation, metabolic dysfunction reciprocally impairing themselves.

Hypoadiponectinemia in sepsis may be set in a picture of adipose tissue dysfunction, called adiposopathy, and characterized not only by accumulation of fat but also by the

* Luca Siracusano viale Regina Margherita 28 98121 Messina Italy 0039 090 46461, Fax 0039 090 2212494 e-mail lsiracusano@unime.it1.

inability of adipose tissue to store correctly lipid giving rise to the lipotoxicity and organ damage. Adiponectin may potentially correct many of the metabolic alterations present in sepsis impacting strongly on outcome. This complex of protective activities may explain the relationships between adiponectin serum levels and the outcome in the septic and critically ill patient.

INTRODUCTION

Despite the novel pathogenetic and therapeutic acquisitions sepsis remains the main cause of death in the critically ill patient (p.); it has been defined as the systemic host reaction to an infection. The typical immune response in sepsis presents a first stage, characterized by excessive production of proinflammatory interleukins (ILs) produced by monocytes, macrophages and endothelial cells in response to a few molecules, common to all virulent pathogens, called pathogen-associated molecular patterns (PAMPs)(1). They are recognized by receptors on the surface of the immune cells, named Pattern Recognition Receptors (PRP) such as Toll-like Receptors (TLR). This phase, named Systemic Inflammatory Reaction Syndrome (SIRS) is followed by a second phase, compensatory anti- inflammatory response syndrome (CARS), that modulates the SIRS. Through this biphasic reaction in the largest part of the cases the homeostasis is reestablished. It is possible however that either during the proinflammatory phase, because of the "cytokine storm"-induced organ damage further increased through the release by injured tissue of the endogenous damage-associated molecular pattern (DAMPs or alarmins) or due to an excessive immunodepression and uncontrollable infections (2), death may ensue. Many therapies aiming at normalizing immune response in sepsis failed until now to improve clinical outcome (3). Adiponectin (APN) an adipokine with antiatherosclerotic, insulin sensitizing and cardioprotective actions has been shown to possess also an anti-inflammatory action and to be reduced in septic patients, its severe deficiency being associated with a higher level of mortality and of proinflammatory cytokines (4). The altered APN secretion in sepsis as well as in critical illness is likely the consequence of adipose tissue dysfunction that may play an important role in the survival of the critical patient. In this article we will stress the relationships among adipose tissue dysfunction and the related alteration of APN and adipokines secretion, the regulation of inflammation and metabolic dysfunction in sepsis and their prognostic influence.

ADIPONECTIN: CHARACTERISTICS AND MECHANISM OF ACTION

The adipose tissue by means of both fat cells and residential macrophages produces many cytokines (IL-6, IL-8, IL-10, TNF-α,) and peptides [APN, leptin, plasminogen activator inhibitor-1 (PAI-1), retinol binding protein 4 (RBP4), resistin, visfatin, chemerin, vaspin and apelin] named adipokines because they are cytokine-like factors produced in the adipose tissue (5). The adiponectin, first identified in 1995 (6), is the adipokine produced in major amount by adipose tissue; circulating levels range from 2 to 30 μg\ml in human, three order of magnitude more than leptin and are more abundant in women than men probably for a depressing effect of testosterone on its production. Its very high plasma level suggests that APN possesses functions not mediated by receptors signaling but involving low affinity macromolecular interactions (7). APN is produced in high amounts only in the adipose tissue even though there are recent evidences of low level release by other types of cells (8-11); its plasma levels are negatively related to body mass index (BMI) and visceral fat accumulation. Hypoadiponectinemia is also present in diabetes, metabolic syndrome, cardiovascular diseases, sepsis and critically ill patients.The cause of APN reduction in obesity and other diseases is not clear but has been showed that proinflammatory cytokines, increased in all these conditions, can downregulate APN expression, moreover it was recently described that this adipose derived hormone behaves as a negative acute phase protein in a dog model of endotoxemia (12). Another explanation for APN downregulation in obesity and type 2 diabetes (T2D) is the reduced level of Sirtuin-1 (SIRT-1) and forkhead box O transcription factor (FOXO-1) both able to bind the promoter of APN increasing its expression (13).

The molecule of human APN is formed by 244 amino acids with an N-terminal signal region, a species-specific variable domain, an intermediate stalk collagen-like domain which is the site undergoing the posttranslational modifications so important for the regulation of its action; finally a carboxyl-terminal globular domain sharing similarities with complement factor 1q and tumor necrosis factor-a (TNF-α)(14).

The APN is encoded by the APM1 gene mapped to chromosome 3q27, a locus associated with susceptibility to metabolic syndrome and type 2 diabetes (T2D) (15). Its transcription is regulated by the interaction of many transcription factors. The promoter contains a PPARγ-responsive element (PPRE) so that the PPARγ agonist thiazolidinediones (TZD) increase APN transcription (16) whereas adipose tissue-specific PPARγ deletion in mice reduces its plasma levels (17). Also the FOXO1, the CCAAT-enhancer-binding proteins (C/EBPα) and the Sterol regulatory element-binding protein (SREBP) may positively regulate APN expression whereas the cAMP response element-binding (CREB), the Nuclear factor of activated T-cells (NFAT) but mainly the proinflammatory ILs TNF-α and IL-6 suppress APN expression in adipocytes (18).

APN is subject to many posttranslational modifications mainly at the level of the collagen-like domain: glycosylation-sialylation regulating clearance, glycosylation and disulphide-bond formation required for multimerisation that affects substantially its action (19). After synthesis the single unit 30 kDa APN is assembled in the endoplasmic reticulum (ER) into higher molecular weight APN (HMWA) and secreted into the circulating form, chiefly trimer, undergoing further polymerization to hexamers and HMW (containing 18-36 molecules). Multimers are much more active on insulin sensitization and multimerization is a mechanism to regulate the different biological actions of adiponectin and to target it to different tissues (20). APN multimerization in the ER has been showed as a complex mechanism involving the molecular chaperones ERp44 (ER protein of 44 kDa) and ER oxidoreductase 1-Lα (Ero1-Lα) both of which are induced during adipogenesis and whose deletion disrupts APN secretion (20): the first blocks APN secretion through a thiol-mediated retention and the second one facilitates its secretion transferring its disulfide bonds to favor APN multimerization. The role played by ER in APN assembly may justify the altered secretion of this adipokine in presence of an ER stress, that is a damage to ER caused by an accumulation of unfolded proteins in pathological conditions such as nutrient excess and obesity (21). A significant demonstration of the importance the posttranslational modifications carry out is the inactivity of recombinant

APN purified from bacteria, devoid of posttranslational modifications, administered to experimental animals whereas the endogenous APN purified from mammals maintains its full activity (22).

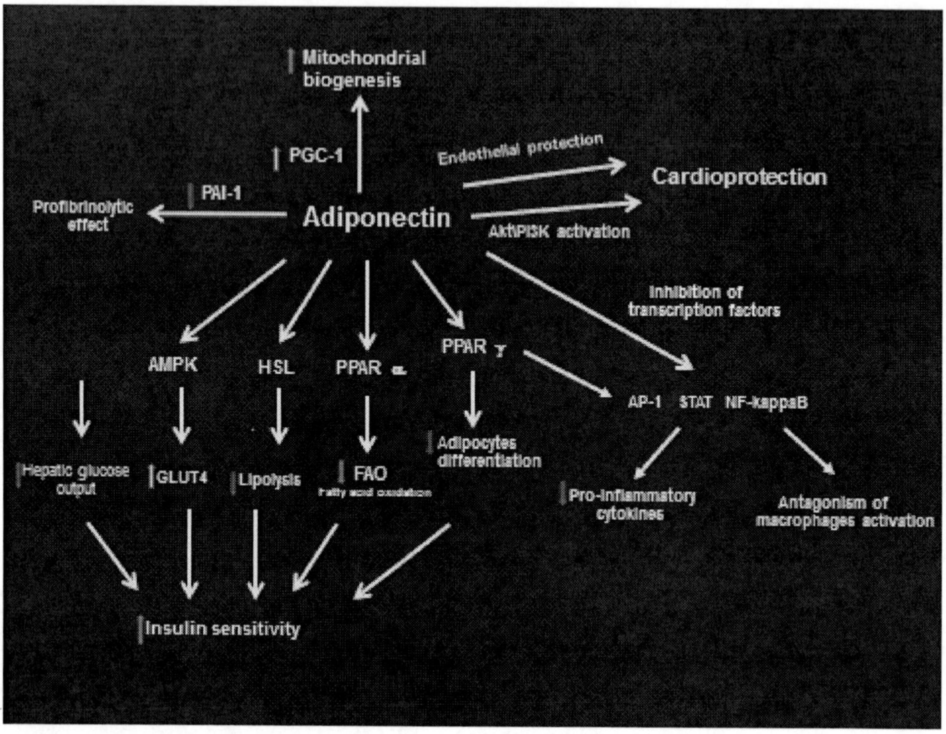

Figure 1. Mechanisms of APN action.

A. RECEPTORS

APN acts through its receptors descovered in 2003 (23): AdipoR1 and AdipoR2. Both receptors are expressed ubiquitously (24,25) but AdipoR1, receptor for globular APN (gAPN), is mainly expressed in skeletal muscle and has been detected in the paraventricular nucleus of hypothalamus where it may be involved in energy regulation (26) while AdipoR2 binds to full-length APN and is chiefly expressed in the liver. These receptors share a seven domains transmembrane structure but on the contrary of G protein coupled receptors the N-terminal domain is intracellular. Also T-cadherin may function as a third receptor for hexameric and full-length APN in vascular endothelial cells, smooth muscle and myocardiocytes; this protein however lacks an intracytoplasmic signalling domain and can only modify the response to AdipoR1-2 competing with them for APN binding. Denzel et colleagues (27) showed that T-cadherin was essential for the cardioprotective action of APN: in an experimental setting of ischemia-reperfusion injury (IRI) in mice the receptor was necessary for the reduction of infarct size and AMPK phosphorylation.

AdipoR1 action is mediated through AMPK activation and APN infusion mimics AMPK stimulation increasing lipid oxidation and translocation of GLUT4, the insulin-dependent membrane glucose transporter, to the plasma membrane in skeletal muscle; the same effects are also present in the liver associated with decreased glucose output due to inhibition of the enzymes of the gluconeogenesis (23). The effect on lipid oxidation is also mediated by AdipoR2 through upregulation of the peroxisome proliferation activating receptors alpha (PPARα), transcription factors regulating the expression of proteins involved in uptake, intracellular transport and utilization of free fatty acids (FFA)(28). Increased lipid oxidation in skeletal muscle seems to require the sequential activation of AMPK, p38 MAPK and PPARα (29). The activation by APN of AMPK as well as PPARα and p38MAPK has been showed in rodents ko or overexpressing AdipoR1-2 (30). In mice overexpressing or ko for AdipoR1 was showed another important action of APN which stimulates Ca^{2+} influx so activating AMPK, the downstream kinase Ca2+/calmodulin-dependent protein kinase kinase β (CAMKKβ), sirtuin1(SIRT1) and the increased expression and decreased acetylation of peroxisome proliferator-activated receptor coactivator-1α (PGC-1α); this last factor increase mitochondrial biogenesis and may contribute to explain the mitochondrial dysfunction and insulin resistance caused by decreased levels of APN (31). Recently an adaptor protein, named adaptor protein containing pleckstrin homology domain phosphotyrosine binding PTB domain and leucine zipper motif (APPL1)(32), has been identified, it is able to interact with the cytoplasmic domain of AdipoR1 and 2 functioning as a scaffolding protein to facilitate APN-stimulated AMPK and p38 MAPK activation (33) and the signaling chain of APN. APPL1 is able to bind to 14 different proteins such as androgen receptor and various signaling molecules with key roles in apoptosis, cell proliferation, chromatin remodeling and cell survival (34). AdipoR2 was found to be induced not only, as already seen, by PPARγ but also by PPARα via activation of APN gene transcription in macrophages (35). PPARγ increase APN action not only increasing its transcription but also at a posttranslational level, their agonists thiazolidinediones (TZD) increase the ratio of high molecular weight to low molecular weight APN by small adipocytes, the upregulation of APN release seems to be an important mediator of PPARγ- induced increase in insulin sensitivity (36-38). The increased production of APN by PPARγ stimulation could also be a major mechanism mediating, almost partly, their antiinflammatory effects that requires an intact adiponectin signaling pathway (39).

A family of new highly conserved proteins with structure and properties similar toAPN, called APN paralogs, has been identified (40,41), it includes osmotin and C1q and TNF-α related proteins (CTRP) and is involved in diverse processes including metabolism, host defense, apoptosis, cell differentiation, and organogenesis (42)

APN and Insulin Sensitivity

The pathogenesis of insulin resistance is no more interpreted on the basis of the glucocentric but of the lipocentric theory (43) stating that the excess of circulating free fatty acids (FFA), the main inhibitors of glucose utilization in the cell, when present in excess in plasma because of increased lipolysis, decreased fat storing capacity of adipose tissue or reduced fatty acid oxydation (FAO) may accumulate in the organs (liver, myocardium, pancreas etc) giving origin to reduced sensitivity to insulin, increased hepatic glucose output and to the phenomenon of lipotoxicity, syndrome characterized by accumulation of FFA in the cell, increased production of ROS, activation of apoptosis (lipoapoptosis) and organ dysfunction (44). APN may interfere with the core of this process increasing, through the stimulation of PPARα, lipid oxidation, reducing hepatic neoglucogenesis through stimulation of AMPK and augmenting glucose uptake in muscle and liver cells by increased translocation of GLUT4 to the plasma membrane (45). APN further favors lipid storage in adipocytes, the differentiation of these cells and their fat storing ability, upregulating PPARγ expression so decreasing the plasma level of FFA and increasing insulin sensitivity. While a normal secretion of APN by healthy adipose cells normally takes place, whenever adipocytes become dysfunctional the production of this adipokine is compromised. The APN levels, mainly HMWA, in man correlate inversely with visceral fat accumulation (46) and directly with insulin sensitivity, the presence of T2D and with the risk to develop diabetes (47,48). This implies that hypoadiponectinemia is related to an adipose tissue dysfunction that may be present in obesity, diabetes, hypertension and in the metabolic syndrome that is a cluster of all them; however hypoadiponectinemia is also present in sepsis and in the critical p. meaning that an adipose tissue dysfunction is also involved in these pathological state.

The function of APN has been studied in APN ko mice that are viable and present severe insulin resistance only if feed a high-fat diet. Also on a normal diet however ko mice display a decreased clearance of plasma FFA, a decreased production of the lipid transport protein fatty acid transport protein 1 (FATP-1) and of TNFα, notoriously linked to insulin resistance; adenoviral mediated APN expression reverses these alterations (45,49). Contrary to T2D APN levels are increased in diabetes type 1 (50) and in the defects of insulin receptors signalling (51) suggesting that insulin may suppress APN secretion and that hypoadiponectinemia in diseased states may be related to hyperinsulinemia (52). A vicious circle between hyperinsulinemia and the downregulation of APN may therefore be established in the initial phases of insulin resistance (53).

APN and Cardioprotection

Hypoadiponectinemia has been considered as a risk factor for cardiovascular diseases and especially for coronary heart disease by some authors (54,55), although more recent articles consider APN levels as a more significant marker for the risk of T2D than for coronary artery disease (CAD) (56,57). Indeed APN may affect positively many of the pathogenetic factors of CAD:

a) inhibiting the TNFα-induced production of the transcription factor NF-kB so blocking the proinflammatory and procoagulant pathways
b) inhibiting the expression of adhesion molecules on macrophages and endothelial cells and inhibiting smooth muscle cells proliferation (58)
c) increasing endothelial nitric oxide production by restoring eNOS function and reactivating high glucose-suppressed endothelial progenitor cell function (59,60)
d) inhibiting smooth muscle cells proliferation and macrophage to foam cell transformation (61)
e) reducing plasma levels of LDL-cholesterol and of oxidized lipoproteins (61).

This cardioprotective effect is probably mediated by all the actions of APN: metabolic, anti-inflammatory, antiapoptotic and antiatherogenic.

Myocardial infarction is associated with a progressive reduction in APN levels inversely correlated with C reactive protein (CRP) levels (62). In myocardial IRI APN is downregulated (63) because of the increased levels of TNFα, the plasma concentration of which, has a reverse relationship with the levels of this adipokine; APN administration may be cardioprotective in this setting through the reduction of oxidative and nitrosative stress induced by the stimulation of both AMPK and Cyclooxygenase-2 (COX-2) activity (64). COX2 has been shown to mediate also the suppression of TNFa signaling through a sphingosine kinase-1_pathway (65) and to inhibit adverse remodeling of myocardium contrasting pathological cardiac growth (66,67). The stimulation of the Reperfusion Injury Salvage Kinase (RISK) pathway through Akt\PI3K and through the inhibition of mitochondrial permeability pore transition (mPTP) [a pore opening in the internal mitochondrial membrane when ROS (reactive oxygen species) and Ca^{2+} accumulate, determining cell death by necrosis or apoptosis] could also be important in the cardioprotective effect of APN (68,69). Furthermore recently it has been shown that APN can be involved in the protection conferred by ischemic preconditioning because its levels are increased in a rat model of repetitive ischemia, reducing the infarct size through an improvement of IRI (70). APN has also an endothelial protective effect promoting phagocytosis of apoptotic cells by binding to calreticulin on the surface of macrophages and forming a bridge between dead cells and them (71). The absence of the protective effect on endothelium caused by APN deficiency markedly impairs outcome in severe sepsis by allowing a diffuse endothelial activation (72).

APN and Inflammation

The anti-inflammatory action may be a key to understand the beneficial effect of APN and the way in which it is depleted in sepsis (1). APN is connected to the inflammation through multiple pathways: there is a reciprocal correlation among the plasma levels of CRP, fibrinogen and APN (73,74). An identical inverse correlation there is between TNFα, IL6 and APN levels (75), conversely APN increases IL-10 and IL-1RA in human leukocytes (76). APN may also downregulate the chemokines expression by macrophages and endothelial cells reducing T-limphocytes recruitment and can induce macrophages polarization towards the M2 phenotype with less production of proinflammatory ILs and increased production of IL-10 (77-78). All these effects seem to be mediated through the suppression in monocytes and macrophages of the toll-like receptors (TLRs) (73), NF-kB and EGR-1 signaling; the two last are transcription factors regulating the expression of many proinflammatory and procoagulant genes (79,80). The block of NF-kB and CRP production is mediated through the AMPK pathway, the inhibition of which, attenuates these effects (81,82). APN may also almost partly mediate the anti-inflammatory action of PPARγ and of their agonist TZDs (83).

APN seems to present also proinflammatory properties in many inflammatory diseases such as rheumatoid arthritis, inflammatory bowel diseases, systemic lupus erythematosus, asthma, allergy and others (84).

APN is depleted in sepsis worsening outcome through an increased proinflammatory cytokines production (4,85), moreover it has been showed able to neutralize lipopolysaccharide (LPS) in a rat model (86). It shares many effects, among which the antiinflammatory one, with carbon monoxide (CO). The administration of CO, through CO-releasing molecules (CO-RMs), or the increase in APN induced by rosiglitazone, both improve outcome in sepsis (87,88). The existence of an axis has therefore been proposed between heme-oxygenase 1 (HO-1), the enzyme producing CO from heme catabolism, and APN able to modulate in sepsis not only inflammation but also mitochondrial biogenesis through an increase in PGC-1 production (89).

Therapeutic Exploitment of APN Activity

APN administration has been extensively used in the experimental setting either with the purified form either with the recombinant one obtained from bacteria and inactive in some settings due to the absence of posttranslational modifications (22). However the therapeutic use is difficult for the choice of the form to administer (gAPN, full length APN, HMWA; purified or recombinant?) for the very short half-life and the high levels of APN that make difficult and very expensive a large-scale production, so the preferred strategy might be to enhance its secretion or effects. The main category of drugs able to modulate APN effects are the TZD (pioglitazone and rosiglitazone) through their effect on PPARγ able to increase, as already seen, APN secretion through transcriptional and posttranscriptional effects but provided also with positive effects on immune system, insulin sensitivity, adipocytes differentiation and lipid metabolism (25,90-92). Also PPARα agonist fibrates may increase APN activity (93) upregulating the expression of AdipoR1-2 in white adipose tissue , a similar PPARα agonist action has also been showed for ω-3 fish oils and statins (94). A modulating action on APN activity has also been attributed to the cannabinoid receptors CB1 antagonist, renin angiotensin inhibitors and SIRT1 activator resveratrol (90-91).

APN in Sepsis and Critically Ill Patient

Despite the vast number of experimental studies in animals, for the most part finding a positive association between APN levels, organ function and outcome, the plasma levels of APN in septic patients have been the subject of few studies not always confirming the experimental data. More numerous the studies performed in critically ill patients, this last topic has recently been reviewed (95). Hillebrand and colleagues (96) in a study in 33 patients with sepsis or septic shock compared them with morbid obese and healthy subjects identifying a specific adipocytokines profile characterized by a marked reduction of APN with a parallel increase in resistin, PAI-1 and proinflammatory cytokines IL-1α, IL-6, IL-8, IL-10 and TNFα and MCP-1.The authors think that these changes may explain the increased insulin resistance in the septic patient as a consequence of the inflammatory state similarly to the one present in the morbid obesity and metabolic syndrome. In a more recent study (97) the same group described the trend of insulin demand in relation to the APN levels in 25 septic p. to study the effects of APN, leptin and resistin on insulin demand. Surprisingly APN levels were positively correlated with insulin demand whereas there was no correlation among insulin demand and leptin or resistin levels. APN levels were substantially normal, they were also not correlated to BMI such as were insulin demand and resistin levels; only leptin showed a positive correlation with BMI. Insulin demand was not even correlated with leptin or resistin. Carried out in a small number of p. this study presents some unexpected results: the APN level is normal in all p. despite high SAPS and SOFA scores, contrary to the expectations APN is not suppressed by insulin administration, nor a correlation does exist among the plasma levels of resistin, leptin and the insulin demand.

In 74 pediatric patients with systemic inflammatory response syndrome (SIRS), sepsis or septic shock, HMWA and resistin were shown to be increased in patients with septic shock on the first day of hospitalization compared with controls and were related to the mortality risk. HMWA and resistin were also increased in survivors (88) but in minor degree. Studies in humans with experimental endotoxemia failed to show alteration of APN level perhaps because of the obvious minor severity of the clinical picture (99,100). Studies in critical patients including also septic patients showed similar results: (101) Venkatesh and colleagues in 23 critically ill p.(8 with sepsis) measured APN levels at 3 and 7 days after admission noticing superimposable values among them but a significant decrease compared to normal subject; a significant relationship was found between plasma cortisol and adiponectin in critically ill patients, both during the early and late phases, whereas an inverse correlation between plasma adiponectin and CRP was present on day 7. The AA. observe that glucocorticoids, inflammation and oxidative stress are known to decrease APN and that cortisol may act through the consensus sequences for glucocorticoid receptor binding in the APN gene (102). Langoush and colleagues studied APN, RBP-4 and leptin in 318 p. with prolonged critical illness of pulmonary origin (103) showing that circulating levels of these adipokines during the acute phase were low, more in septic than in nonseptic and in p. subjected to acute surgical stress. With time the adipokines increased except APN that remained low only in septic p. Intensive Insulin therapy (IIT) increaseded APN, blunted the rise of RBP4, and did not alter leptin levels. The AA. state that acute stress transiently reduce APN, RBP-4 and leptin and that APN and RBP-4 increase after IIT are markers of improved insulin sensitivity induced by therapy. Koch and colleagues (104) studied in 170 critically ill p. (122 with sepsis) APN levels upon admission to ICU observing no difference between p. (septic vs non septic) and compared with healthy controls. Only p. with preexisting obesity or diabetes had reduced plasma levels of APN. The AA. infer that APN levels in critically ill p. are mainly correlated to insulin resistance. They might not detect any correlation among APN and IL-6, IL-10, TNFα nor the biomarkers of inflammation CRP and procalcitonin PCT). Low levels of APN at admission to ICU are an independent predictor of survival in their experience. Walkey and colleagues (105) in 175 p. with acute respiratory failure enrolled in EDEN pilot study (about 30% with sepsis or pneumonia) retrospectively observed that APN measured within 48 h from the onset of respiratory failure (APN1) was inversely correlated with BMI and higher in females than males; moreover higher APN1 was significantly associated with 28-day mortality. APN levels increased significantly from day 1 to day 6 only in survivors but APN levels at day 6 did not correlate to mortality. The serum APN did not correlate with levels of circulating proinflammatory cytokines TNFα, interferon-gamma, IL1-β, IL-6, IL-8, IL-12 but was associated with the anti-inflammatory cytokine IL-10. In a comment to this study Dembinski (106) claims that in the early phase of critical illness lower

APN levels are a consequence of the severity of illness whereas in the later phase an increased APN level possibly improves convalescence. We think it is possible that the increased levels of APN in the later phase may be expression, such as it is the improvement in mitochondrial biogenesis (107), of the getting through the acute cytokine storm and the consequent downregulation of APN secretion.

Altogether human studies gave results not always reconciliable with experimental studies; many may be the possible explanations such as co-pathologies, small number of p. studied, concomitant therapies, differences in humans and animal response etc.

Dysfunction of Adipose Tissue and Adipokines Secretion in Sepsis

Obesity is not always associated with an increased risk of death in p. with critical illness and cardiovascular diseases (CVD) and mortality seems to be paradoxically reduced in overweight (BMI between 25 and 30 kg/m^2) and obese (BMI between 30 and 40 kg/m^2) (108-115) but not in morbide obese (BMI $\geq$ 40 kg/m^2) (116,117) ICU p. although there is not a general agreement about this topic (118-120). Particularly interesting are the results of the study by Paolini and colleagues (121) showing that a high sagittal abdominal diameter, not BMI, is an independent risk factor of death in critically ill patients . The concept of the relative protection conferred by obesity, named “obesity paradox”, was first observed in hemodialysis and cardiac p. (122,123) and may give rise to many questions: in which way obesity can modify the outcome in critical illness and interfere with organ function? are the alterations observed in sepsis caused by pure obesity or by a dysfunction of adipose tissue not strictly related to BMI? The response may come from the recent acquisitions about adipose tissue pathophysiology.

The adipose tissue is the largest organ of the body in humans and is no more regarded as an inert deposit of lipid but it is now considered important for metabolic and cardiovascular healthy, for the immune function and for determination of lifetime length (124-127). The adipose tissue undergoes dysfunction during sepsis and due to the profoundly modified cross-talk with other organs may contribute to their impairment through many mechanisms. As caloric intake and BMI increase in an individual leading to obesity, adipose tissue shows characteristic modifications: it becomes infiltrated by macrophages and inflamed, adipocytes become hypertrophic producing less APN and more proinflammatory cytokines and adipokines (resistin and RBP4) reducing insulin sensitivity. Adipocytes hypertrophy is however only a temporary mean to store the excess of calories as fat but it causes a contemporary stimulus to the recruitment, proliferation and differentiation of other adipocytes (128). On the contrary of what was believed the adipocytes number is not invariable and these cells may increase their number renewing annually by 10% (129). So the number of adipocytes increase parallel to a temporary hypertrophy of adipocytes. If however adipocytes differentiation, controlled mainly by PPARγ and SREBP-1c, is impaired and fails, the hypertrophy of adipocytes goes on, the process of storing lipid in adipose tissue is compromised and FFA, main inhibitors of glucose utilization, begin to accumulate in vital organs giving origin to reduced insulin sensitivity, increased hepatic glucose output and to lipotoxicity with impaired cardiac, pancreatic and liver function. In this way obesity can impact on the function of the body increasing the tendency towards hyperglycemia, hyperlipidemia, reduced insulin sensitivity, reduced APN secretion, increased production of resistin and RBP4 and inflammation because of increased proinflammatory cytokines production in adipose tissue that are the components of the metabolic syndrome, a risk factor for cardiovascular diseases (CVD) but also for immune system alterations prominent in sepsis. This process of compromised differentiation of preadipocytes leads to a picture similar to the lipoatrophy, a condition of reduced ability of adipose tissue to store lipid, complicated by organ accumulation of FFA and lipotoxicity and it has been named adiposopathy namely adipose tissue anatomical and functional abnormalities promoted by positive caloric balance resulting in metabolic, endocrine and immune derangements and to increased CVD incidence (130). Moreover the inability of subcutaneous fat to store FFA in excess bypasses them to the visceral fat, more prone to inflammation, lipolysis and production of contrainsular adipokines further increasing organ damage. The ability to face and manage an increased load of FFA may differentiate obese subjects from p. with adiposopathy, prone to develop lipotoxicity (128). The decreased availability of PPARα and PPARγ and the reduced FAO observed in sepsis (see paragraph on APN and metabolic dysfunction in sepsis), potentially correctible by APN, may explain lipotoxicity observed in sepsis . Especially the lipid infiltration of liver seems to be a marker of adipose tissue pathology (131). The increased risk in sepsis as well as in critical illness would thus be related to adiposopathy and not to the pure increase of fat mass.

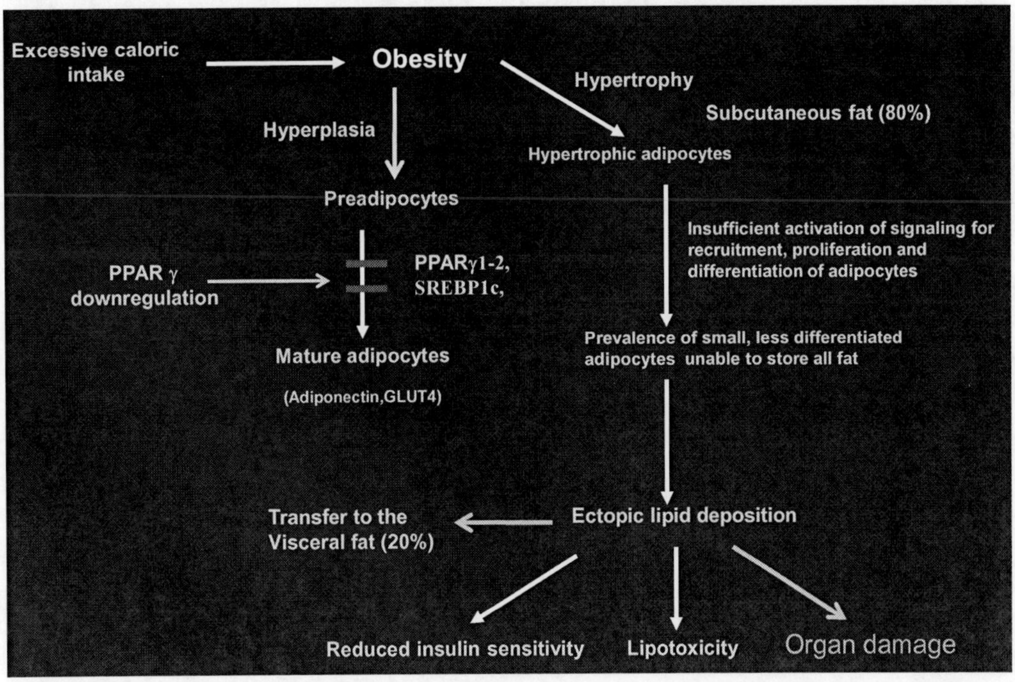

Figure 2. Mechanism of adiposopathy establishment.

According to some new studies carried out mainly by McLaughlin and colleagues in humans and experimental animals, in diabetes and insulin-resistent states there is a prevalence in the adipose tissue of small adipocytes representing a population less differentiated (132-

135), with decreased expression of differentiation markers and enhanced inflammatory activity (decreased PPARγ and APN) (136-137). Moreover the failure to store fat in subcutaneous adipose tissue leads to store it in visceral adipose tissue with the already seen negative consequences; TZD may correct the failure of differentiation and reduce the visceral fat accumulation increasing the deposit in subcutaneous fat (138,139). The increased abdominal storage of fat (male obesity) is a sign of adiposopathy because it behaves as visceral fat correlating to the increased risk as observed in the study by Paolini (121) and is probably caused by the presence of more androgen receptors in visceral fat.

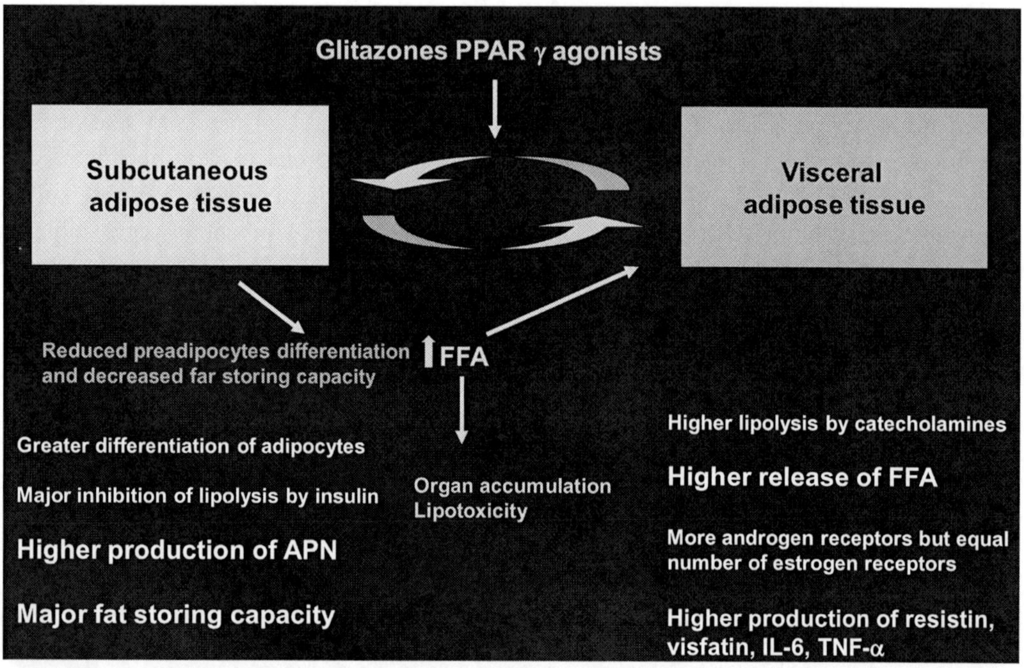

Figure 3. Differential regulation and roles of subcutaneous and visceral fat.

The only study on the role of adipose tissue in critically ill p. is the bright and innovative article by Langouche and colleagues (140); in a large prospective randomized study on critically ill p. randomly assigned to conventional or intensive insulin therapy (see paper for details) they performed post mortem adipose biopsy in 61 critically ill p. and 20 matched controls. Adipose morphology was also studied in in vivo biopsies of 27 patients and in a rabbit model of critical illness. Adipose tissue from critically ill p. showed a higher number of macrophages and of smaller adipocytes with increased levels of GLUT1-3 and glucose and a reduced level of GLUT4. The smaller adipocytes presented an increased presence of the marker Preadipocyte Factor-1 (Pref-1). The AA interpreted the results as expression of an increased ability of small adipocytes to take up circulating glucose and triglycerides. We could provide an alternative explanation: although the novel view by McLaughlin is not accepted by all, if it were true, the accumulation of smaller cells in the adipose tissue could not be an adaptive and protective response but a sign of impairment of the function of the adipose tissue during the acute phase of sepsis contributing to worsen the endocrine, metabolic and immune functions, the more in p. with a full-blown sepsis. On the other hand the increased GLUT1-3, particularly the isoform 1 that is the stress and proinflammatory cytokines-responsive, with reduced insulin-sensitive GLUT4 could be due to the inflammatory stress known to increase them. This point of view might be further corroborated by the positivity of the smaller cells for Preadipocyte Factor-1 (Pref-1), an inhibitor of preadipocytes differentiation (141,142), that might be explained as a proliferation not followed by differentiation of the preadipocytes.

APN and Metabolic Dysfunction in Sepsis

The experimental studies, mainly performed in mice, demonstrated the ability of APN to correct in many ways the metabolic alterations present in sepsis and correlated to the outcome (143). We think therefore it may be useful to review some metabolic pathways profoundly modified in sepsis and possible target for a therapeutic correction by APN.

The etiology of insulin resistance in sepsis is related to the metabolic storm involving energy metabolism with increased proteic catabolism regarding mainly muscle and parenchymatous organs and relatively preserving adipose tissue (140,144). Energy consumption is increased in the early phases of sepsis, under the influence of the proinflammatory ILs, but tends to be reduced in the later ones because of the hypoperfusion (144). The glucose metabolism is severely compromised because of the increase of the contrainsular substances (glucagon, TNFalpha, growth hormone, cortisol, FFA) with augmented neoglucogenesis, hyperglycemia, hyperlactatemia and reduced insulin sensitivity at the muscular and hepatic level (144). The endogenous insulin secretion in the critically ill p. is increased, on the contrary of what was suggested in the past (145) and remains so also during IIT. Insulin is provided wiyh an antiinflammatory action that is missing if a reduced sensitivity to its action is present such as in critical illness further impairing inflammation (146).

Indeed, hyperglycemia causes an increased mitochondrial oxidation of glucose and consequently an increased production of ROS mainly in non insulin-dependent cells in which the membrane transport of glucose is ensured, even in conditions of relative insulin shortage, by ubiquitous GLUT1, whose production is increased by cytokines, while insulin-dependent cells are relatively protected because the lack of insulin reduces the availability of GLUT4 (147). The disproportionate increase of ROS production is worsened because of the concomitant mitochondrial dysfunction and the reduction in scavengers, typical of the critical patient. Thus, one can understand the ability of insulin to control glucose oxidation and to reduce oxidative stress and organ dysfunction in animal models (147). Lipolysis is increased during sepsis (148) because of the increased activity of the lipases of the adipose tissue, mainly the hormone sensitive lipase (HSL), activated by catecholamines and TNFα and inhibited by APN (149). An overload of FFA and of very low density lipoproteins (VLDL) is thus induced along with a reduction in heavy density lipoproteins (HDL), which have been attributed with a positive role in sepsis for their immunomodulant action and their ability to bind and neutralize endotoxin (150).

Feingold and colleagues in a series of studies (151-153) showed that LPS administration in mice determines a reduction of the transcription receptors PPARα and γ and of PGC-1α, as well as of Thyroid receptors α and β (TRα and TRβ), Liver X receptors α (LXRα), and of Retinoid X receptors (RXR) in the myocardium, diaphragm and kidneys. All these transcription factors are involved in the expression of proteins critical for lipid metabolism that is therefore severely impaired in sepsis. The simultaneous downregulation of the two isoforms of PPAR may have a deleterious effect in sepsis because PPARα deficiency reduces FAO while the cells are flooded by a load of FFA due to the augmented lipolysis. By consequence FFA because of reduced ability of subcutaneous adipose tissue to store them are transferred to visceral fat or accumulated in vital organs damaging their function through lipotoxicity . On the other hand the downregulation of PPARγ inhibits the differentiation of the subcutaneous adipose cells giving rise to the reduced ability of adipose tissue to store fat and to the adiposopathy.

The importance of PPARα in sepsis is also highlighted by their ability to reduce the expression of inflammatory ILs and to induce apoptosis of macrophages, T lymphocytes, neutrophils and immunocompetent cells in general, so reducing inflammation (154), and to protect the heart from IRI (155). The control of the immune response is moreover controlled by PPARγ, able to inhibit the transcription factors AP-1, STAT and NF-kB through a mechanism of transrepression, so reducing the production of the proinflammatory cytokines and the activation of the macrophages. The action of PPARγ may be useful during SIRS to modulate the inflammatory response but may impair outcome during CARS leading to diffuse infection (154).

What is the role of APN in this intricate picture? Many experimental researches described the increased mortality in APN-KO mice subjected to CLP-induced sepsis that has been attributed to the increased production of proinflammatory ILs (4,72,87,88) . On the metabolic front APN through stimulation of AMPK and PPARα may increase FAO and reduce lipolysis preventing lipotoxicity (149) , may furthermore increase PGC-1 and mitochondrial biogenesis whose reactivation is linked to survival in sepsis (107), antagonizing LPS, reducing hepatic glucose output and restoring insulin sensitivity. The function of the adipose tissue and the normalization of adipokines secretion may be improved through the stimulation of PPARγ, with a positive resultant effect on many of the pathophysiological alterations described in sepsis.

Conclusion

APN has, in our opinion, the potential to correct many metabolic derangements present in sepsis and its global effect is protective in sepsis and critical illness as confirmed by the experimental but also from the few available human studies. APN action is strictly linked to the adipose tissue function and undergoes modifications together with it and the adipokines network. It is probable that in the near future we can understand better how the dysfunction of adipose tissue and consequently of adipokines production may impact , through the cross-talk with other organs, on the global function of the body and the outcome of sepsis.

References

[1] Cinel I, Opal SM: Molecular biology of inflammation and sepsis: a primer. *Crit Care Med.* 2009 Jan;37(1):291-304.

[2] Opal SM: New perspectives on immunomodulatory therapy for bacteraemia and sepsis. *Int J Antimicrob Agents*. 2010 Dec;36 Suppl 2:S70-3.

[3] Antonopoulou A, Giamarellos-Bourboulis EJ: Immunomodulation in sepsis: state of the art and future perspective. *Immunotherapy*. 2011 Jan;3(1):117-28.

[4] Uji Y, Yamamoto H, Tsuchihashi H, Maeda K, Funahashi T, Shimomura I, Shimizu T, Endo Y, Tani T: Adiponectin deficiency is associated with severe polymicrobial sepsis, high inflammatory cytokine levels, and high mortality. *Surgery*. 2009 May;145(5):550-7.

[5] Galic S, Oakhill JS, Steinberg GR: Adipose tissue as an endocrine organ. *Mol Cell Endocrinol*. 2010;316(2):129-39.

[6] Scherer PE, Williams S, Fogliano M, Baldini G, Lodish HF: A novel serum protein similar to C1q, produced exclusively in adipocytes. *J Biol Chem.* 1995 Nov 10;270(45):26746-9.

[7] Ouchi N, Walsh K: A novel role for adiponectin in the regulation of inflammation. *Arterioscler Thromb Vasc Biol*. 2008 Jul;28(7):1219-21.

[8] Takano H, Obata JE, Kodama Y, Kitta Y, Nakamura T, Mende A, Kawabata K, Saito Y, Fujioka D, Kobayashi T, Yano T, Sano K, Kugiyama K: Adiponectin is released from the heart in patients with heart failure. *Int J Cardiol.* 2009 Feb 20;132(2):221-6.

[9] Delaigle AM, Jonas JC, Bauche IB, Cornu O, Brichard SM: Induction of adiponectin in skeletal muscle by inflammatory cytokines: in vivo and in vitro studies. *Endocrinology* 2004; 145: 5589–5597.

[10] Berner HS, Lyngstadaas SP, Spahr A et al. Adiponectin and its receptors are expressed in bone-forming cells. *Bone* 2004; 35: 842–849.

[11] Nishida M, Funahashi T, Shimomura I: Pathophysiological significance of adiponectin. *Med Mol Morphol* 2007; 40: 55-67.

[12] Tvarijonaviciute A, Eralp O, Kocaturk M, Yilmaz Z, Ceron JJ: Adiponectin and IGF-1 are negative acute phase proteins in a dog model of acute endotoxaemia. *Vet Immunol Immunopathol*. 2011 Mar 15;140(1-2):147-51.

[13] Qiao L, Shao J: SIRT1 regulates adiponectin gene expression through Foxo1-C/enhancer-binding protein alpha transcriptional complex. *J Biol Chem.* 2006 Dec 29;281(52):39915-24.

[14] Guerre-Millo M: Adiponectin: an update. *Diabetes Metab*. 2008 Feb;34(1):12-8.

[15] Kissebah AH, Sonnenberg GE, Myklebust J et al: Quantitative trait loci on chromosomes 3 and 17 influence phenotypes of the metabolic syndrome. *Proc Natl Acad Sci USA 2000*; 97: 14478–14483

[16] Yu JG, Javorschi S, Hevener AL, Kruszynska YT, Norman RA, Sinha, M.Olefsky JM: (2002) The effect of thiazolidinediones on plasma adiponectin levels in normal, obese, and type 2 diabetic subjects. *Diabetes* 51, 2968–2974.

[17] He W, Barak Y, Hevener A, Olson P, Liao D, Le J, Nelson , Ong E, OlefskyJM, Evans RM: Adipose-specific peroxisome proliferator-activated receptor γ knockout causes insulin resistance in fat and liver but not in muscle. Proc. Natl. Acad. Sci. U.S.A. *2003*;100, 15712–15717.

[18] Liu M, Liu F: Transcriptional and post-translational regulation of adiponectin. *Biochem J*. 2009 Dec 14;425(1):41-52.

[19] 19)Simpson F, Whitehead JP: Adiponectin-it's all about the modifications. *The International Journal of Biochemistry & Cell Biology* 2010, 42:785-788.

[20] Wang Y, Lam KS, Yau MH, Xu A: Post-translational modifications of adiponectin: mechanisms and functional implications. *Biochem J.* 2008 Feb 1;409(3):623-33.

[21] Gregor MF, Hotamisligil GS: Thematic review series: adipocyte biology. Adipocyte stress: the endoplasmic reticulum andmetabolic disease. *J Lipid Res* 2007, 48:1905-1914.

[22] Wang Y, Lu G, Wong WP et al. Proteomic and functional characterization of endogenous adiponectin purified from fetal bovine serum. *Proteomics* 2004; 4: 3933–3942.

[23] Yamauchi T, Kamon J, Ito Y, Tsuchida A, Yokomizo T, Kita S, Sugiyama T, Miyagishi M, Hara K, Tsunoda M, Murakami K, Ohteki T, Uchida S, Takekawa S, Waki H, Tsuno NH, Shibata Y, Terauchi Y, Froguel P, Tobe K, Koyasu S, Taira K, Kitamura T, Shimizu T, Nagai R, Kadowaki T: Cloning of adiponectin receptors that mediate antidiabetic metabolic effects. *Nature*. 2003 Jun 12;423(6941):762-9.

[24] Kadowaki T, Yamauchi T, Kubota N, Hara K, Ueki K, Tobe K: Adiponectin and adiponectin receptors in insulin resistance, diabetes, and the metabolic syndrome. *J Clin Invest.* 2006 Jul;116(7):1784-92.

[25] Kadowaki T, Yamauchi T: Adiponectin and adiponectin receptors. *Endocr Rev.* 2005 May;26(3):439-51.

[26] Kubota N, Yano W, Kubota T, et al. Adiponectin stimulates AMPactivated protein kinase in the hypothalamus and increases food intake. *Cell Metab* 2007;6:55–68.

[27] Denzel MS, Scimia MC, Zumstein PM, Walsh K, Ruiz-Lozano P, Ranscht B: T-cadherin is critical for adiponectin-mediated cardioprotection in mice. *J Clin Invest.* 2010 Dec 1;120(12):4342-52. 18)

[28] Lopaschuk GD, Ussher JR, Folmes CD, Jaswal JS, Stanley WC: Myocardial fatty acid metabolism in health and disease. *Physiol Rev.* 2010 Jan;90(1):207-58.

[29] Yoon MJ, Lee GY, Chung JJ, Ahn YH, Hong SH, Kim JB: Adiponectin increases fatty acid oxidation in skeletal muscle cells by sequential activation of AMP-activated protein kinase, p38 mitogen-activated protein kinase, and peroxisome proliferator-activated receptor alpha. *Diabetes*. 2006 Sep;55(9):2562-70.

[30] Whitehead JP, Richards AA, Hickman IJ, Macdonald GA & Prins JB (2006). Adiponectin – a key adipokine in the metabolic syndrome. *Diabetes Obes Metab* 8, 264–280.

[31] Iwabu M, Yamauchi T, Okada-Iwabu M, Sato K, Nakagawa T, Funata M, Yamaguchi M, Namiki S, Nakayama R, Tabata M, Ogata H, Kubota N, Takamoto I, Hayashi YK, Yamauchi N, Waki H, Fukayama M, Nishino I, Tokuyama K, Ueki K, Oike Y, Ishii S, Hirose K, Shimizu T, Touhara K, Kadowaki T: Adiponectin and AdipoR1 regulate PGC-1alpha and mitochondria by Ca(2+) and AMPK/SIRT1. *Nature* 2010;464, 1313–1319.

[32] Mao X, Kikani CK, Riojas RA, Langlais P, Wang L, Ramos FJ, et al: APPL1 binds to adiponectin receptors and mediates adiponectin signaling and function. *Nat Cell Biol* 2006;8:516–23.

[33] Xin X, Zhou L, Reyes CM, Liu F, Dong LQ: APPL1 mediates adiponectin-stimulated p38 MAPK activation by scaffolding the TAK1-MKK3-p38 MAPK pathway. *Am J Physiol Endocrinol Metab.* 2011 Jan;300(1):E103-10.

[34] Brochu-Gaudreau K, Rehfeldt C, Blouin R, Bordignon V, Murphy BD, Palin MF: Adiponectin action from head to toe. *Endocrine*. 2010 Feb;37(1):11-32.

[35] Chinetti G, Zawadski C, Fruchart JC, Staels B: Expression of adiponectin receptors in human macrophages and regulation by agonists of the nuclear receptors PPARalpha, PPARgamma, and LXR. *Biochem Biophys Res Commun.* 2004 Jan 30;314(1):151-8.

[36] Pajvani, U. B., M. Hawkins, T. P. Combs, M. W. Rajala, T. Doebber, J. P. Berger, J. A. Wagner, M. Wu, A. Knopps, A. H. Xiang, K. M. Utzschneider, S. E. Kahn, J. M. Olefsky, T. A. Buchanan, and P. E. Scherer. 2004. Complex distribution, not absolute amount of adiponectin, correlates with thiazolidinedione-mediated improvement in insulin sensitivity. *J. Biol. Chem.* 279:12152–12162.

[37] Bouskila M, Pajvani UB, Scherer PE: Adiponectin: a relevant player in PPARgamma-agonist-mediated improvements in hepatic insulin sensitivity? *Int J Obes* (Lond). 2005 Mar;29 Suppl 1:S17-23.

[38] Sharma AM, Staels B: Review: Peroxisome proliferator-activated receptor gamma and adipose tissue understanding obesity-related changes in regulation of lipid and glucose metabolism. *J Clin Endocrinol Metab.* 2007 Feb;92(2):386-95.

[39] Aprahamian T, Bonegio RG, Richez C, Yasuda K, Chiang LK, Sato K, Walsh K, Rifkin IR: The peroxisome proliferator-activated receptor gamma agonist rosiglitazone ameliorates murine lupus by induction of adiponectin. *J Immunol.* 2009 Jan 1;182(1):340-6.

[40] Wong GW, Wang J, Hug C, Tsao TS, Lodish HF. A familyof Acrp30/adiponectin structural and functional paralogs. *Proc Natl Acad Sci U S A 2004*;101:10302–10307.

[41] Narasimhan ML, Coca MA, Jin J, Yamauchi T, Ito Y,Kadowaki T, Kim KK, Pardo JM, Damsz B, Hasegawa PM, et al: Osmotin is a homolog of mammalian adiponectin and controls apoptosis in yeast through a homolog of mammalian adiponectin receptor. *Mol Cell* 2005;17:171–180.

[42] Kim MJ, Lee W, Park EJ, Park SY: C1qTNF-related protein-6 increases the expression of interleukin-10 in macrophages. *Mol Cells.* 2010 Jul;30(1):59-64.

[43] Mittra S, Bansal VS, Bhatnagar PK: From a glucocentric to a lipocentric approach towards metabolic syndrome. *Drug Discov Today.* 2008 Mar;13(5-6):211-8.

[44] Unger RH, Clark GO, Scherer PE, Orci L: Lipid homeostasis, lipotoxicity and the metabolic syndrome.) *Biochim Biophys Acta.* 2010 Mar;1801(3):209-14.

[45] Okamoto Y, Kihara S, Funahashi T, Matsuzawa Y, Libby P: Adiponectin: a key adipocytokine in metabolic syndrome. *Clin Sci* (Lond). 2006 Mar;110(3):267-78.

[46] Matsuzawa Y, Funahashi T, Nakamura T: The Concept of Metabolic Syndrome: Contribution of Visceral Fat Accumulation and Its Molecular Mechanism. *J Atheroscler Thromb.* 2011 Jul 8.

[47] Kern PA, Di Gregorio GB, Lu T, Rassouli N, Ranganathan G: Adiponectin expression from human adipose tissue: relation to obesity, insulin resistance, and tumor necrosis factor-alpha expression. *Diabetes* 2003,52:1779-1785.

[48] Samuel VT, Petersen KF, Shulman GI: Lipid-induced insulin resistance: unravelling the mechanism. *Lancet.* 2010 Jun 26;375(9733):2267-77.

[49] Maeda N, Shimomura I, Kishida K et al: Diet-induced insulin resistance in mice lacking adiponectin/ACRP30. *Nat Med* 2002; 8: 731–737.

[50] Imagawa A, Funahashi T, Nakamura T, Moriwaki M, Tanaka S, NishizawaH, SayamaK, UnoS, IwahashiH,YamagataK, Miyagawa J, Matsuzawa Y: Elevated serum concentration of adipose-derived factor, adiponectin, in patients with type 1 diabetes. *Diabetes Care* 2002; 25:1665–1666.

[51] Semple RK, Soos MA, Luan J, Mitchell CS, Wilson JC, Gurnell M, Cochran EK, Gorden P, Chatterjee VK, Wareham NJ, O'Rahilly S Elevated plasma adiponectin in humans with genetically defective insulin receptors. *J Clin Endocrinol Metab* 2006; 91:3219–3223

[52] Cook JR, Semple RK: Hypoadiponectinemia--cause or consequence of human "insulin resistance"? *J Clin Endocrinol Metab.* 2010 Apr;95(4):1544-54.

[53] Basu R, Pajvani UB, Rizza RA, Scherer PE Selective downregulation of the high molecular weight form of adiponectin in hyperinsulinemia and in type 2 diabetes: differential regulation from nondiabetic subjects. *Diabetes* 2007;56:2174–2177.

[54] Kumada M, Kihara S, Sumitsuji S, Kawamoto T, Matsumoto S, Ouchi N, Arita Y, Okamoto Y, Shimomura I, Hiraoka H, Nakamura T, Funahashi T, Matsuzawa Y. Association of hypoadiponectinemia with coronary artery disease in men. *Arterioscler Thromb Vasc Biol* 23: 85–89, 2003.

[55] Pischon T, Girman CJ, Hotamisligil GS et al. Plasma adiponectin levels and risk of myocardial infarction in men. *JAMA* 2004;291:1730–1737.

[56] Wannamethee SG, Tchernova J, Whincup P, et al: Associations of adiponectin with metabolic and vascular risk parameters in the British Regional Heart Study reveal stronger links to insulin resistance-related than to coronary heart disease risk-related parameters. *Int J Obes* (Lond). 2007;31(7):1089–1098.

[57] Lawlor DA, Davey Smith G, Ebrahim S, et al: Plasma adiponectin levels are associated with insulin resistance, but do not predict future risk of coronary heart disease in women. *J Clin Endocrinol Metab.* 2005;90(10):5677–5683.

[58] Maia-Fernandes T, Roncon-Albuquerque R Jr, Leite-Moreira AF: Cardiovascular actions of adiponectin: pathophysiologic implications. *Rev Port Cardiol.* 2008 Nov;27(11):1431-49.

[59] Chen H, Montagnani M, Funahashi T, Shimomura I, Quon MJ: Adiponectin stimulates production of nitric oxide in vascular endothelial cells. *J Biol Chem.* 2003 Nov 7;278(45):45021-6.

[60] Huang PH, Chen JS, Tsai HY, Chen YH, Lin FY, Leu HB, Wu TC, Lin SJ, Chen JW: Globular adiponectin improves high glucose-suppressed endothelial progenitor cell function through endothelial nitric oxide synthase dependent mechanisms. *J Mol Cell Cardiol.* 2011 Jul;51(1):109-19.

[61] Barseghian A, Gawande D, Bajaj M: Adiponectin and vulnerable atherosclerotic plaques. *J Am Coll Cardiol.* 2011 Feb 15;57(7):761-70.

[62] Kojima S, Funahashi T, Sakamoto T, et al: The variation of plasma concentrations of a novel, adipocyte derived protein, adiponectin, in patients with acute myocardial infarction. *Heart* 2003;89:667.

[63] Liu S, Yin T, Wei X, Yi W, Qu Y, Liu Y, Wang R, Lian K, Xia C, Pei H, Sun L, Ma Y, Lau WB, Gao E, Koch WJ, Wang H, Tao L: Downregulation of adiponectin induced by tumor necrosis factor α is involved in the aggravation of posttraumatic myocardial ischemia/reperfusion injury. *Crit Care Med.* 2011 Apr 14. [Epub ahead of print]

[64] Shibata R, Sato K, Pimentel DR, Takemura Y, Kihara S, Ohashi K, Funahashi T, Ouchi N, Walsh K: Adiponectin protects against myocardial ischemia-reperfusion injury through AMP kinase- and COX-2-dependent mechanisms. *Nat Med* 2005;11:1096–1103.

[65] Ikeda Y, et al. Cyclooxygenase-2 induction by adiponectin is regulated by a sphingosine kinase-1 dependent mechanism in cardiac myocytes. *FEBS Lett* 2008;582:1147–1150.

[66] Goldstein BJ, Scalia RG, Ma XL: Protective vascular and myocardial effects of adiponectin. *Nat Clin Pract Cardiovasc Med.* 2009 Jan;6(1):27-35.

[67] Ouchi N, Shibata R, Walsh K: Cardioprotection by adiponectin. *Trends Cardiovasc Med.* 2006 Jul;16(5):141-6.

[68] Gonon AT, Widegren U, Bulhak A, Salehzadeh F, Persson J, Sjöquist PO, Pernow J: Adiponectin protects against myocardial ischaemia-reperfusion injury via AMP-activated protein kinase, Akt, and nitric oxide. *Cardiovasc Res.* 2008 Apr 1;78(1):116-22.

[69] Smith CC, Yellon DM: Adipocytokines, cardiovascular pathophysiology and myocardial protection. *Pharmacol Ther.* 2011 Feb;129(2):206-19.

[70] Wang H, Cheng JS, Wu WJ, Wen JY, Yu CA, Liao WQ, Kong W, Ke YN, Zheng JG: Ischemia preconditioning attenuated myocardial ischemia via upregulating the expression of adiponectin in rat*]. Zhonghua Xin Xue Guan Bing Za Zhi.* 2010 Oct;38(10):929-33 (english abstract).

[71] Takemura Y, Ouchi N, Shibata R, Aprahamian T, Kirber MT, Summer RS, Kihara S, Walsh K. Adiponectin modulates inflammatory reactions via calreticulin receptor-dependent clearance of early apoptotic bodies. *J Clin Invest.* 2007;117:375–386.

[72] Teoh H, Quan A, Bang KW, Wang G, Lovren F, Vu V, Haitsma JJ, Szmitko PE, Al-Omran M, Wang CH, Gupta M, Peterson MD, Zhang H, Chan L, Freedman J, Sweeney G, Verma S: Adiponectin deficiency promotes endothelial activation and profoundly exacerbates sepsis-related mortality. *Am J Physiol Endocrinol Metab* 2008, 295:E658-664.

[73] Ouchi N, Kihara S, Funahashi T et al. Reciprocal association of C-reactive protein with adiponectin in blood stream and adipose tissue. *Circulation* 2003; 107:671–674.

[74] Schulze MB, Rimm EB, Shai I, Rifai N, Hu FB: Relationship between adiponectin and glycemic control, blood lipids, and inflammatory markers in men with type 2 diabetes. *Diabetes Care* 2004; 27: 1680–1687.

[75] Kern PA, Di Gregorio GB, Lu T, Rassouli N, Ranganathan G: Adiponectin expression from human adipose tissue: relation to obesity, insulin resistance, and tumor necrosis factor-alpha expression. *Diabetes* 2003,52:1779-1785.

[76] Okamoto Y, Folco EJ, Minami M, Wara AKMK, Feinberg MW, Sukhova GK, Colvin RA, Kihara S, Funahashi T, Luster AD, Libby P. Adiponectin inhibits the production of CXC receptor 3 chemokine ligands in macrophages and reduces T-lymphocyte recruitment in atherogenesis. *Circ Res.* 2008;102:218-225.

[77] K. Ohashi, J. L. Parker, N. Ouchi et al., "Adiponectin promotes macrophage polarization toward an anti-inflammatory phenotype," *The Journal of Biological Chemistry,* 285, 9, pp. 6153–6160, 2010.

[78] Yamaguchi, N. et al. Adiponectin inhibits Toll-like receptor family-induced signaling. *FEBS Lett.* 579,6821–6826 (2005).

[79] Wolf AM, Wolf D, Rumpold H, Enrich B, Tilg H: Adiponectin induces the anti-inflammatory cytokines IL-10 and IL-1RA in human leukocytes. *Biochem Biophys Res Commun.* 2004 Oct 15;323(2):630-5.

[80] Park PH, McMullen MR, Huang H, Thakur V, Nagy LE: Short-term treatment of RAW264.7 macrophages with adiponectin increases tumor necrosis factor-alpha (TNF-alpha) expression via ERK1/2 activation and Egr-1 expression: role of TNF-alpha in adiponectin-stimulated interleukin-10 production. *J Biol Chem.* 2007 Jul 27;282(30):21695-703.
[81] Ouchi N, Kihara S, Arita Y et al. Adiponectin, an adipocyte-derived plasma protein, inhibits endothelial NF-kappaB signaling through a cAMP-dependent pathway. *Circulation* 2000; 102: 1296–1301.
[82] Devaraj S, Torok N, Dasu MR, Samols D, Jialal I: Adiponectin decreases C-reactive protein synthesis and secretion from endothelial cells: evidence for an adipose tissue-vascular loop. *Arterioscler Thromb Vasc Biol.* 2008 Jul;28(7):1368-74.
[83] Ouchi N, Walsh K: Adiponectin as an anti-inflammatory factor. *Clin Chim Acta.* 2007 May 1;380(1-2):24-30.
[84] Fantuzzi, G. Adiponectin and inflammation: consensus and controversy. *J.Allergy Clin. Immunol.* 121, 326–330 (2008).
[85] Lancel S, Hassoun SM, Favory R, Decoster B, Motterlini R, Neviere R: Carbon monoxide rescues mice from lethal sepsis by supporting mitochondrial energetic metabolism and activating mitochondrial biogenesis. *J Pharmacol Exp Ther.* 2009 May;329(2):641-8.
[86] Tsuchihashi H, Yamamoto H, Maeda K, Ugi S, Mori T, Shimizu T, Endo Y, Hanasawa K, Tani T: Circulating concentrations of adiponectin, an endogenous lipopolysaccharide neutralizing protein, decrease in rats with polymicrobial sepsis. *J Surg Res.* 2006 Aug;134(2):348-53.
[87] Uji Y, Yamamoto H, Maeda K, Tsuchihashi H, Akabori H, Shimizu T, Endo Y, Shimomura I, Tani T: Adiponectin deficiency promotes the production of inflammatory mediators while severely exacerbating hepatic injury in mice with polymicrobial sepsis. *J Surg Res.* 2010 Jun 15;161(2):301-11.
[88] Cepinskas G, Katada K, Bihari A, Potter RF: Carbon monoxide liberated from carbon monoxide-releasing molecule CORM-2 attenuates inflammation in the liver of septic mice. *Am J Physiol Gastrointest Liver Physiol.* 2008 Jan;294(1):G184-91.
[89] Siracusano L, Girasole V: Carbon monoxide and adiponectin in sepsis. *Surgery.* 2010 May;147(5):755.
[90] Sun Y, Xun K, Wang C, Zhao H, Bi H, Chen X, Wang Y: Adiponectin, an unlocking adipocytokine. *Cardiovasc Ther* 2009;27: 59–75.
[91] Phillips SA, Kung JT: Mechanisms of adiponectin regulation and use as a pharmacological target. *Curr Opin Pharmacol.* 2010 Dec;10(6):676-83.
[92] Shetty S, Kusminski CM, Scherer PE: Adiponectin in health and disease: evaluation of adiponectin-targeted drug development strategies. *Trends Pharmacol Sci* 2009, 30:234-239.
[93] Tsuchida A, Yamauchi T, Takekawa S, Hada Y, Ito Y, Maki T, Kadowaki T: Peroxisome proliferator-activated receptor (PPAR)alpha activation increases adiponectin receptors and reduces obesity-related inflammation in adipose tissue: comparison of activation of PPARalpha, PPARgamma, and their combination. *Diabetes.* 2005 Dec;54(12):3358-70.
[94] Perelas A, Tsoulkani A, Perrea D: Effects of lipid-lowering drugs on adiponectin. *Curr Vasc Pharmacol.* 2010 Nov;8(6):836-48.

[95] Robinson K, Prins J, Venkatesh B: Clinical review: Adiponectin biology and its role in inflammation and critical illness. *Crit Care*. 2011 Apr 20;15(2):221.

[96] 96)Hillenbrand A, Knippschild U, Weiss M, Schrezenmeier H, Henne-Bruns D, Huber-Lang M, Wolf AM: Sepsis induced changes of adipokines and cytokines - septic patients compared to morbidly obese patients. *BMC Surg*. 2010 Sep 9;10:26.

[97] Hillenbrand A, Weiss M, Knippschild U, Stromeyer HG, Henne-Bruns D, Huber-Lang M, Wolf AM. Association of adiponectin levels and insulin demand in critically ill patients. *Diabetes Metab Syndr Obes*. 2011 Jan 26;4:45-51.

[98] Kaplan JM, Denenberg A, Monaco M, Nowell M, Wong H, Zingarelli B: Changes in peroxisome proliferator-activated receptor-gamma activity in children with septic shock. *Intensive Care Med* 2010, 36:123-130.

[99] Keller P, Moller K, Krabbe KS, Pedersen BK (2003) Circulating adiponectin levels during human endotoxaemia. *Clin Exp Immunol* 134:107–110.

[100] Anderson PD, Mehta NN, Wolfe ML,Hinkle CC, Pruscino L, Comiskey LL, Tabita-Martinez J, Sellers KF, Rickels MR, Ahima RS, Reilly MP (2007) Innate immunity modulates adipokines in humans. *J Clin Endocrinol Metab* 92:2272–2279.

[101] Venkatesh B, Hickman I, Nisbet J, Cohen J, Prins J: Changes in serum adiponectin concentrations in critical illness: a preliminary investigation. *Crit Care*. 2009;13(4):R105.

[102] Comuzzie AG, Funahashi T, Sonnenberg G, Martin LJ, Jacob HJ, Black AE, Mass D, Takahashi M, Kihara S: The genetic basis of plasma variation in adiponectin, a global endophenotype for obesity and the metabolic syndrome. *J Clin Endocrinol Metab* 2001, 86:4321-4325.

[103] Langouche L, Vander Perre S, Frystyk J, Flyvbjerg A, Hansen TK, Van den Berghe G: Adiponectin, retinol-binding protein 4, and leptin in protracted critical illness of pulmonary origin. *Crit Care* 2009, 13:R112.

[104] Koch A, Sanson E, Voigt S, Helm A, Trautwein C, Tacke F: Serum adiponectin upon admission to the intensive care unit may predict mortality in critically ill patients. *J Crit Care*. 2011 Apr;26(2):166-74.

[105] Walkey AJ, Rice TW, Konter J, Ouchi N, Shibata R, Walsh K, deBoisblanc BP, Summer R: Plasma adiponectin and mortality in critically ill subjects with acute respiratory failure. *Crit Care Med*. 2010 Dec;38(12):2329-34.

[106] Dembinski R: Adiponectin in critically ill patients: more questions than answers? *Crit Care Med*. 2010 Dec;38(12):2415-6.

[107] Carré JE, Orban JC, Re L, Felsmann K, Iffert W, Bauer M, Suliman HB, Piantadosi CA, Mayhew TM, Breen P, Stotz M, Singer M: Survival in critical illness is associated with early activation of mitochondrial biogenesis. *Am J Respir Crit Care Med*. 2010 Sep 15;182(6):745-51.

[108] Childers DK, Allison DB. The 'obesity paradox': a parsimonious explanation for relations among obesity, mortality rate and aging? *International Journal of Obesity* (Lond) 2010;34:1231–8.

[109] Peake SL, Moran JL, Ghelani DR, Lloyd AJ, Walker MJ: The effect of obesity on 12-month survival following admission to intensive care: a prospective study. *Crit Care Med* 2006;34:2929-2939.

[110] Nasraway SA Jr, Albert M, Donnelly AM, Ruthazer R, Shikora SA, Saltzman E: Morbid obesity is an independent determinant of death among surgical critically ill patients. *Crit Care Med* 2006;34:964-970.

[111] Amundson DE, Djurkovic S, Matwiyoff GN: The obesity paradox. *Crit Care Clin.* 2010 Oct;26(4):583-96.

[112] Vachharajani V: Influence of obesity on sepsis. *Pathophysiology.* 2008; Aug; 15(2):123-34.

[113] Garrouste-Orgeas M, Troché G, Azoulay E, Caubel A, de Lassence A, Cheval C, Montesino L, Thuong M, Vincent F, Cohen Y, Timsit JF: Body mass index. An additional prognostic factor in ICU patients. *Intensive Care Med.* 2004Mar ; 30(3):437-43.

[114] Lewandowski K, Lewandowski M: Intensive care in the obese. *Best Pract Res Clin Anaesthesiol.* 2011 Mar;25(1):95-108.

[115] Goulenok C, Monchi M, Chiche JD, Mira JP, Dhainaut JF, Cariou A: Influence of overweight on ICU mortality: a prospective study. Chest. 2004 Apr;125(4):1441-5. *Chest.* 2005 Jun;127(6):2125-31.

[116] Akinnusi ME, Pineda LA, El Solh AA: Effect of obesity on intensive care morbidity and mortality: a meta-analysis. *Crit Care Med.* 2008 Jan;36(1):151-8.

[117] Prabhakar G, Haan CK, Peterson ED, Coombs LP, Cruzzavala JL, Murray GF: The risks of moderate and extreme obesity for coronary artery bypass grafting outcomes: a study from the Society of Thoracic Surgeons' database. *Ann Thorac Surg.* 2002 Oct;74(4):1125-30.

[118] Ray DE, Matchett SC, Baker K, Wasser T, Young MJ: The effect of body mass index on patient outcomes in a medical ICU. *Chest.* 2005 Jun;127(6):2125-31.

[119] Aldawood A, Arabi Y, Dabbagh O: Association of obesity with increased mortality in the critically ill patient. *Anaesth Intensive Care.* 2006 Oct;34(5):629-33.

[120] Sakr Y, Madl C, Filipescu D, et al. Obesity is associated with increased morbidity but not mortality in critically ill patients. *Intensive Care Med* 2008;34:1999-2009.

[121] Paolini JB, Mancini J, Genestal M, Gonzalez H, McKay RE, Samii K, Fourcade OA: Predictive value of abdominal obesity vs. body mass index for determining risk of intensive care unit mortality. *Crit Care Med.* 2010 May;38(5):1308-14.

[122] Kalantar-Zadeh K, Block G, Humphreys MH, Kopple JD: Reverse epidemiology of cardiovascular risk factors in maintenance dialysis patients. *Kidney Int.* 2003 Mar;63(3):793-808.

[123] Kalantar-Zadeh K, Block G, Horwich T, Fonarow GC: Reverse epidemiology of conventional cardiovascular risk factors in patients with chronic heart failure. *J Am Coll Cardiol.* 2004 Apr 21; 43(8):1439-44.

[124] Galic S, Oakhill JS, Steinberg GR: Adipose tissue as an endocrine organ. *Mol Cell Endocrinol.* 2010 Mar 25;316(2):129-39.

[125] Tchkonia T, Morbeck DE, Von Zglinicki T, Van Deursen J, Lustgarten J, Scrable H, Khosla S, Jensen MD, Kirkland JL: Fat tissue, aging, and cellular senescence. *Aging Cell.* 2010 Oct;9(5):667-84.

[126] Arai Y, Takayama M, Abe Y, Hirose N: Adipokines and Aging. *J Atheroscler Thromb.* 2011 May 7. [Epub ahead of print].

[127] Kaminski DA, Randall TD: Adaptive immunity and adipose tissue biology. *Trends Immunol.* 2010 Oct;31(10):384-90.

[128] Bays HE: Adiposopathy is "sick fat" a cardiovascular disease? *J Am Coll Cardiol*. 2011 Jun 21;57(25):2461-73.

[129] Spalding KL, Arner E, Westermark PO, et al. Dynamics of fat cell turnover in humans. *Nature* 2008;453:783–7.

[130] Bays HE, Gonzalez-Campoy JM, Bray GA, et al. Pathogenic potential of adipose tissue and metabolic consequences of adipocyte hypertrophy and increased visceral adiposity. *Expert Rev Cardiovasc Ther* 2008;6:343– 68.

[131] Magkos F, Fabbrini E, Mohammed BS, et al. Increased whole-body adiposity without a concomitant increase in liver fat is not associated with augmented metabolic dysfunction. *Obesity* (Silver Spring) 2010;18:1510–5.

[132] McLaughlin T, Sherman A, Tsao P, et al. Enhanced proportion of small adipose cells in insulin-resistant vs insulin-sensitive obese individuals. *Diabetologia* 2007;50:1707–15.

[133] Liu A, Sonmez A, Yee G, Bazuine M, Arroyo M, Sherman A, McLaughlin T, Reaven G, Cushman S, Tsao P: Differential adipogenic and inflammatory properties of small adipocytes in Zucker Obese and Lean rats. *Diab Vasc Dis Res*. 2010 Oct;7(4):311-8.

[134] Kursawe R, Eszlinger M, Narayan D, Liu T, Bazuine M, Cali AM, Adamo ED, Shaw M, Pierpont BM, Shulman GI, Cushman SW, Sherman A, Caprio S: Cellularity and adipogenic profile of the abdominal subcutaneous adipose tissue from obese adolescents: association with insulin resistance and hepatic steatosis. *Diabetes* 2010;59:2288–2296.

[135] McLaughlin T, Deng A, Yee G, Lamendola C, Reaven G, Tsao PS, Cushman SW, Sherman A: Inflammation in subcutaneous adipose tissue: relationship to adipose cell size. *Diabetologia*. 2010 Feb;53(2):369-77.

[136] Dubois SG, Heilbronn LK, Smith SR, et al. Decreased expression of adipogenic genes in obese subjects with type 2 diabetes. *Obesity* (Silver Spring) 2006;14:1543–52

[137] Yang X, Jansson PA, Nagaev I, Jack MM, Carvalho E, Sunnerhagen KS, Cam MC, Cushman SW, Smith U. Evidence of impaired adipogenesis in insulin resistance. *Biochem Biophys Res Commun*. 2004; 317: 1045–1051.

[138] McLaughlin TM, Liu T, Yee G, Abbasi F, Lamendola C, Reaven GM, Tsao P, Cushman SW, Sherman A: Pioglitazone increases the proportion of small cells in human abdominal subcutaneous adipose tissue. *Obesity* (Silver Spring). 2010 May;18(5):926-31.

[139] Miyazaki Y, Mahankali A, Matsuda M, et al: Effect of pioglitazone on abdominal fat distribution and insulin sensitivity in type 2 diabetic patients. *J Clin Endocrinol Metab* 2002;87:2784-91.

[140] Langouche L, Perre SV, Thiessen S, Gunst J, Hermans G, D'Hoore A, Kola B, Korbonits M, Van den Berghe G: Alterations in adipose tissue during critical illness: An adaptive and protective response? *Am J Respir Crit Care Med*. 2010 Aug.

[141] Kim KA, Kim JH, Wang Y, Sul HS: Pref-1 (preadipocyte factor 1) activates the MEK/extracellular signal-regulated kinase pathway to inhibit adipocyte differentiation. *Mol Cell Biol* 2007;27:2294-2308.

[142] Wang Y, Kim KA, Kim JH, Sul HS: Pref-1, a preadipocyte secreted factor that inhibits adipogenesis. *J Nutr*. 2006 Dec;136(12):2953-6.

[143] Van Cromphaut SJ, Vanhorebeek I, Van den Berghe G: Glucose metabolism and insulin resistance in sepsis. *Curr Pharm Des*. 2008;14(19):1887-99.

[144] Tappy L, Chioléro R: Substrate utilization in sepsis and multiple organ failure. *Crit Care Med* 2007; 35[Suppl.]:S531–S534.
[145] 1Duska F, Andel M: Intensive blood glucose control in acute and prolonged critical illness: endogenous secretion contributes more to plasma insulin than exogenous insulin infusion. *Metabolism*. 2008 May;57(5):669-71.
[146] Li L, Messina JL: Acute insulin resistance following injury. *Trends Endocrinol Metab*. 2009 Nov;20(9):429-35.
[147] Van den Berghe G: How does blood glucose control with insulin save lives in intensive care? *J Clin Invest* 2004; 114:1187–1195.
[148] Khovidhunkit W, Kim MS, Memon RA, Shigenaga JK, Moser AH, Feingold KR, Grunfeld C: Effects of infection and inflammation on lipid and lipoprotein metabolism: mechanisms and consequences to the host. *J Lipid Res*. 2004 Jul;45(7):1169-96.
[149] Qiao L, Kinney B, Schaack J, Shao J: Adiponectin inhibits lipolysis in mouse adipocytes. *Diabetes*. 2011 May;60(5):1519-27.
[150] Murch O, Collin M, Hinds CJ, Thiemermann C: Lipoproteins in inflammation and sepsis.I. Basic science. *Intensive Care Med* 2007; 33:13–24.
[151] Feingold K, Kim MS, Shigenaga J, Moser A, Grunfeld C: Altered expression of nuclear hormone receptors and coactivators in mouse heart during the acute-phase response. *Am J Physiol Endocrinol Metab*. 2004;286(2):E201-7.
[152] Feingold KR, Wang Y, Moser A, Shigenaga JK, Grunfeld C: LPS decreases fatty acid oxidation and nuclear hormone receptors in the kidney. *J Lipid Res*.2008; 49: 2179-87.
[153] Feingold KR, Moser A, Patzek SM, Shigenaga JK, Grunfeld C: Infection decreases fatty acid oxidation and nuclear hormone receptors in the diaphragm. *J Lipid* Res. 2009;50:2055-63.
[154] von Knethen A, Soller M, Brüne B: Peroxisome proliferator-activated receptor gamma (PPAR gamma) and sepsis. *Arch Immunol Ther Exp* 2007; 55:19-25.
[155] Abdelrahman M, Sivarajah A, Thiemermann C.: Beneficial effects of PPARgamma ligands in ischemia-reperfusion injury, inflammation and shock. *CardiovascRes* 2005; 65:772-81

In: Adiponectin: Production, Regulation and Roles in Disease ISBN: 978-1-61942-165-3
Editor: Masayoshi Yamaguchi

Chapter VII

HYPERLIPIDEMIA INDUCED IN REGUCALCIN TRANSGENIC RATS: INVOLVEMENT OF ADIPONECTIN

Masayoshi Yamaguchi [1]
Division of Endocrinology and Metabolism and Lipids,
Department of Medicine,
Emory University School of Medicine,
WMRB, Atlanta, Georgia, US

ABSTRACT

Regucalcin is known as a regulatory protein in various intracellular signaling systems. Regucalcin transgenic (TG) rat was generated to determine the role of regucalcin in metabolic disorders. Regucalcin TG rat has been found to induce hyperlipidemia with increasing age (14 to 50 weeks); serum triglyceride, high-density lipoprotein (HDL)-cholesterol, free fatty acid, albumin, and calcium concentrations are markedly increased in the TG male and female rats with increasing age. The decrease in lipid and glycogen contents in the liver tissues was induced in regucalcin TG rats. The expression of leptin mRNA in the adipose and liver tissues was decreased in regucalcin TG rats of 50 weeks old. Adiponectin mRNA levels were not changed in the adipose tissues of regucalcin TG rats of 50 weeks old, while its expression was decreased in the liver tissues. Leptin and adiponectin are adipokines that are involved in lipid metabolism. These decreases may be partly involved in the hyperlipidemia induced in regucalcin TG rats. Moreover, overexpression of regucalcin has been shown to enhance glucose utilization and lipid production in the cloned rat hepatoma H4-II-E cells *in vitro*, and insulin resistance is seen in the cells. The expression of glucose transporter 2 mRNA was increased in the transfectants which overexpress regucalcin, while the expressions of insulin receptor and phosphatidylinositol 3-kinase mRNAs that are involved in insulin signaling. It is proposed that regucalcin is related to hyperlipidemia and that the

[1] E-mail address: yamamasa1155@yahoo.co.jp

regucalcin TG rat model may be useful to determine the pathophysiological state and the development of therapeutic tool for hyperlipidemia.

Keywords: regucalcin, RGN, lipid, hyperlipidemia, insulin registance, adiponectin, leptine

INTRODUCTION

Calcium ion (Ca^{2+}) plays an important role in the regulation of many cell functions. The Ca^{2+} effect on cells is amplified by Ca^{2+}-dependent protein kinases that relate to a signal transduction due to hormonal stimulation [1- 3]. Regucalcin was found in the year 1978 as a novel Ca^{2+}-binding protein that differs from calmodulin and other Ca^{2+}-related proteins as it does not contain an EF-hand motif of Ca^{2+}-binding domain [4-7]. The name regucalcin was proposed for this Ca^{2+}-binding protein that may regulate the effect of Ca^{2+} on cell functions [8-12]. There is growing evidence that regucalcin plays a multifunctional role as a regulatory protein in Ca^{2+} dependent and independent signaling processes in many cell types (reviewed in references 13-20).

Regucalcin and its gene (*Rgn*) have been identified in 16 species comprising of regucalcin family [7, 21, 22]. Comparison of the nucleotide sequences of regucalcin from vertebrate species is highly conserved in their coding region, and they conserve throughout evolution [21, 22]. The gene is localized on the X chromosome in rat and human [23, 24]. The organization of rat regucalcin gene consists of seven exons and six introns, and several consensus regulatory elements exist upstream of the 5'-flanking region [25]. AP-1 [26], NFI-A1 [27, 28], RGPR-p117 [29-31], and β-catenin [32] have been found to be a transcription factor for the enhancement of regucalcin gene promoter activity. The transcription activity of the regucalcin gene is enhanced through various intracellular signaling factors. This enhancement is partly involved in the phosphorylation of nuclear proteins that mediate through various protein kinases related to intracellular signaling [33-38].

Regucalcin greatly expresses in the liver and kidney cortex [39-41]. Regucalcin is also found in other tissues and many cells including brain, heart, bone, breast, lung, and prostate [42-46], suggesting that the protein has a role in the regulation of their tissue functions. The expression of regucalcin mRNA is regulated by the administration of various factors including calcium [41], insulin [47], 17β-estradiol, and androgen [45, 46, 48] in the tissues of rats *in vivo*.

Regucalcin has a multifunctional role in maintaining intracellular Ca^{2+} homeostasis by activating Ca^{2+} pump enzymes; suppressing Ca^{2+} signaling from the cytoplasm to the nucleus in the proliferating cells; inhibiting protein kinases, protein phosphatases, protein synthesis, deoxyribonucleic acid (DNA) and ribonucleic acid (RNA) synthesis in the cytoplasm and nucleus of cells and suppressing apoptotic cell death induced by various signaling factors (reviewed in references 15-20). From these findings, regucalcin has been proposed to play a pivotal role in maintaining cell homeostasis and function as the regulatory protein of intracellular signaling system [20].

The physiological role of regucalcin *in vivo* has been poorly understood, however. The deficiency of regucalcin (RGN/SMP30) induces a decrease in L-ascorbic acid (vitamin C) in mice [49]. This may not be significant, because vitamin C is not synthesized in humans. We studied a role of endogenous regucalcin *in vivo* using a transgenic (TG) rat model [50]. The great expression of endogenous regucalcin was demonstrated in the various tissues of regucalcin TG rats [50]. The regulatory effect of endogenous regucalcin on various enzyme activities that are seen by using normal rat tissues *in vitro* [51] has also been demonstrated in regucalcin TG rats *in vivo* [50]. Regucalcin TG rat was a good model to determine the role of endogenous regucalcin with overexpression in the body functions. We found that osteoporosis and hyperlipidemia are induced in regucalcin TG rats, indicating a pathophysiological role of regucalcin in metabolic disorder [44, 52].

This chapter has been written to outline the recent advances that have been made concerning the mechanism by which hyperlipidemia is induced in regucalcin TG rats, and it is suggested an involvement of adiponectin.

Generation of Regucalcin Transgenic Rats

Regucalcin TG rat was generated by pronuclear injection of the transgene with a cDNA construct encoding rat regucalcin [50]. TG founders were fertile, transmitted the transgene at the expected frequency, and bred to homozygote. Western analysis of the cytosol prepared from the tissue from TG female rats (5 weeks old) showed a remarkable expression of regucalcin (33 kDa) protein in the liver, kidney cortex, heart, lung, stomach, brain, spleen, muscle, colon, duodenum, and others [50]. The expression of regucalcin in TG male rats was seen in the liver, kidney cortex, heart, and lung. In wild type (male and female rats), regucalcin mainly was present in the liver and kidney cortex [50]. Sexual differences were seen in the tissue expression of regucalcin in the TG rats [50].

Regucalcin TG rat is registered and preserved in The National BioResource Project for the Rat in Japan (NBRP Rat No:0166, Institute of Laboratory Animals, Graduate School of Medicine, Kyoto University, Kyoto, Japan), and it can get commercially from Japan SLC (Hamamatsu, Japan; http://www.jslc.co.jp/).

Regucalcin has been shown to have a suppressive effect on protein phosphatase activity in the cytosol of liver and kidney cortex from wild-type rats [51]. Dephosphorylation of many phosphorylated proteins is regulated by protein phosphatase in cells [52]. The effect of regucalcin in suppressing protein phosphatase activity was enhanced in the cytosol of kidney cortex of TG male and female rats as compared with those of wild-type rats [50]. Such an effect was associated with the enhancement of regucalcin expression in the TG rats. The decrease in protein phosphatase activity in the TG rats was completely abolished in the presence of anti-regucalcin monoclonal antibody in the enzyme reaction mixture *in vitro* [50]. These observations demonstrate that the endogenous regucalcin has a suppressive effect on protein phosphatase activity *in vivo*.

Regucalcin has been shown to increase Ca^{2+}-ATPase activity in the microsomes (sarcoplasmic reticulum) of rat heart muscle [43]. The Ca^{2+} current is one of the most important components in cardiac excitation-contraction coupling. This coupling mechanism is based on the regulation of intracellular Ca^{2+} concentration by Ca^{2+} pump in the sarcoplasmic reticulum of heart muscle. Regucalcin TG male and female rats caused a remarkable increase in regucalcin expression in the cytosol and microsomes of heart muscle [50]. The microsomal Ca^{2+}-ATPase activity was raised in the heart muscle of the TG male and female rats [50]. This may support the view that endogenous regucalcin has a role as an activator for Ca^{2+}-

ATPase in the microsomes of rat heart muscle *in vivo*. The increase in heart muscle microsomal Ca^{2+}-ATPase activity that was seen in the TG female rats was completely abolished after the addition of anti-regucalcin monoclonal antibody [50]. This indicates that the change in the enzyme's activity in the tissues of regucalcin TG rats is related to an increase in endogenous regucalcin.

The body weight of regucalcin TG female rats was significantly lowered as compared with that of wild-type rats [50]. Serum inorganic phosphorus concentration was increased in TG male and female rats, while serum glucose, triglyceride, free cholestrerol, albumin, and urea nitrogen concentrations were not altered in the TG rats (5 weeks old) [50]. These observations suggest that a specific metabolic disorder, which relates to regulation of serum inorganic phosphorus concentration, is induced in regucalcin TG weanling rats.

Thus, regucalcin TG rat has been shown to be a useful model to define a role of endogenous regucalcin in various tissues *in vivo*.

OVEREXPRESSION OF REGUCALCIN AND HYPERLIPIDEMIA *IN VIVO*

The endogenous regucalcin in the TG rats had a regulatory effect on hepatic nuclear protein phosphatase activity [53], renal cortex cytosolic NO synthase activity [54], heart sarcoplasmic reticulum Ca^{2+}-ATPase and cytosolic NO synthase [43, 55], and brain cytosolic NO synthase and protein tyrosine phosphatase activities [56] *in vivo*. These observations coincided with their regulatory effects of exogenous regucalcin *in vitro* using the subcellular fraction of various tissues obtained from normal rats. Thus, the physiological role of regucalcin in the regulation of cellular function has been demonstrated *in vivo* using regucalcin TG rats. In the aspect of pathophysiological state, regucalcin TG rat was found to induce osteoporosis [44], which is a phenotype in the TG rats.

Interestingly, hyperlipidemia was also induced in regucalcin TG rats with increasing age (14, 25, 36, or 50 weeks) [52]. Serum triglyceride, HDL-cholesterol and free fatty acid concentrations were markedly increased in regucalcin TG male and female rats at 25, 36, or 50 weeks of age [52]. The change in serum HDL-cholesterol concentration in regucalcin TG rats with increasing age is shown in Figure 1.

Serum albumin concentration was also elevated in regucalcin TG female rats at 25, 36, or 50 weeks of age [53]. Moreover, serum calcium concentration was elevated in regucalcin TG male and female rats at 50 weeks of age [52]. Serum zinc, glucose, or urine nitrogen concentrations were not altered in TG male and female rats [52]. Hyperlipidemia was observed from the early stage with increasing age in regucalcin TG rats. Hyperlipidemia that is induced in regucalcin TG rats with increasing age may be unique [52]. Regucalcin TG rat may be useful as an animal model for hyperlipidemia associated with osteoporosis.

Hyperlipidemia has been shown to induce in the lipoprotein lipase-deficient mice [57], low-density lipoprotein (LDL) receptor-deficient mice [58], apolipoprotein C3-knockout mice [59], apolipoprotein C1 transgenic mice [60], very low-density lipoprotein receptor knockout mice [61], cholesterol 7 alpha-hydroxylase-deficient mice [62], apoE-deficient mice [63], and hepatic myr-Akt overexpressing mice [64].

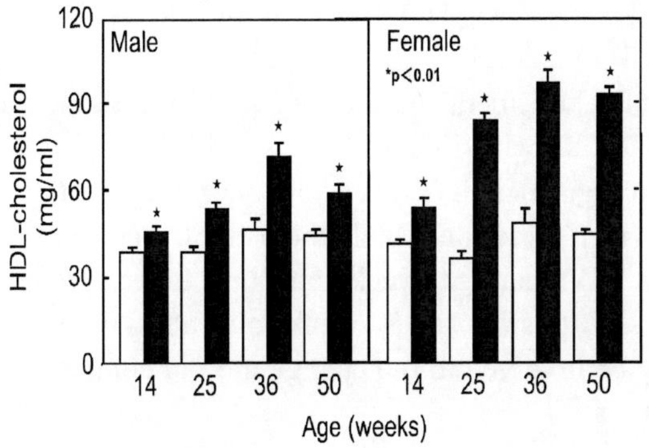

Figure 1. Change in serum HDL-cholesterol concentration in regucalcin transgenic (TG) rats with increasing age. Serum HDL-cholesterol concentration siginificantly ($p<0.01$) increased in the TG rats as compared with that of wild-type rats.

These animal models for hyperlipidemia are involved in molecules that regulate lipid metabolism. Regucalcin may be a novel molecule that participates in lipid metabolism.

Regucalcin is greatly expressed in the liver of rats [39, 40]. Regucalcin has a suppressive effect on the activations of particulate glycogen phosphorylase *a* [65], liver cytosolic pyruvate kinase [66], and fructose 1,6-diphosphatase [67] by Ca^{2+}/calmodulin in the liver of rats, indicating that regucalcin plays a role in the regulation of liver glucose metabolism. Regucalcin may have a regulatory effect on glucose metabolism in the liver. It is assumed that hyperlipidemia is related to production of lipids in the liver of regucalcin TG rats. The role of regucalcin in the regulation of hepatic lipid metabolism has been poorly understood, however.

Regucalcin TG rat also caused a significant increase in serum albumin concentration with increasing age [52]. Overexpression of regucalcin may stimulate production of albumin from the liver of the TG rats. The pathophysiological significance which regucalcin TG rat causes an increase in serum albumin concentration remains to be elucidated.

Serum calcium concentration was significantly raised in regucalcin TG rats at 50 weeks of age [52]. Hypercalcemia may be related to bone loss that is induced in regucalcin TG rats. A significant increase in serum inorganic phosphorus concentration was seen in regucalcin TG rats of 5, 14, 25, or 35 weeks old, but it was not seen in the TG rats of 50 weeks old [52]. Bone loss was induced in regucalcin TG rats with those ages. Hyperphosphataemia and hypercalcemia may also be involved in bone loss in the TG rats.

Thus, regucalcin TG rats in female rather than male have been shown to induce various metabolic disorders, suggesting a pathophysiological importance. Regucalcin TG rat may be useful as animal model in the study of experimental metabolic diseases.

OVEREXPRESSION OF REGUCALCIN AND LIPID METABOLISM *IN VIVO*

Whether lipid components in the adipose and liver tissues are changed in regucalcin TG rats *in vivo* was examined [68]. Regucalcin TG female rats were used at 7 or 50 weeks old. Serum triglyceride or HDL-cholesterol concentrations were significantly increased in regucalcin TG rats of 7 weeks old as compared with those in normal rats of the same age [68].

Serum triglyceride, total cholesterol, HDL-cholesterol, and free fatty acid concentrations were increased in regucalcin TG rats of 50 weeks old [68].
The disorder of lipid metabolism in the adipose and liver tissues, moreover, was found to induce in regucalcin TG rats with increasing age [68]. Triglyceride content in the adipose tissues was increased in regucalcin TG rats of 50 weeks old, while the free fatty acid content was not changed. Liver triglyceride, total cholesterol, and free fatty acid contents were decreased in regucalcin TG rats of 50 weeks old [68]. Liver glycogen content was markedly decreased in regucalcin TG rats of 7 or 50 weeks old [68]. Hyperlipidemia that is induced in regucalcin TG rats may be involved in the change in lipid components in the adipose and live tissues of regucalcin TG rats.

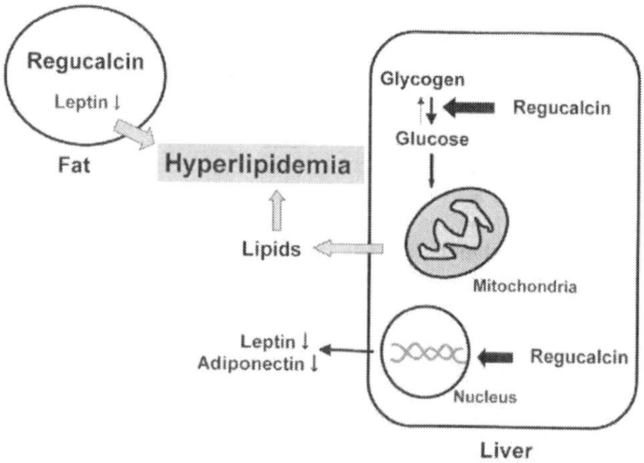

Figure 2. The disorder of lipid metabolism in the adipose and liver tissues is induced in regucalcin transgenic (TG) rats with increasing age. Triglyceride content in the adipose tissues is increased in regucalcin TG rats, while triglyceride, total cholesterol and free fatty acid contents in the liver tissues are decreased in the TG rats. Hyperlipidemia that is induced in regucalcin TG rat may be resulted from the increase in lipid components in the liver tissues and the suppression of leptin or adiponectin mRNA expression in adipose or liver tissues of regucalcin TG rats.

Regucalcin mRNA and its protein were expressed in the adipose tissues of normal rats. However, regucalcin expression in the adipose tissues was not increased in the TG rats [68]. Whether regucalcin in the adipose tissues stimulates the release of lipid components into the serum in regucalcin TG rats is unknown. Triglyceride content in the adipose tissues was increased in regucalcin TG rats [68]. Presumably, the increase in serum lipid components may partly result from the release of lipids from the adipose tissues.

Serum triglyceride concentration was increased in regucalcin TG rats of 7 weeks old as compared with that of normal rats [68]. Triglyceride contents in the adipose and liver tissues were not changed in regucalcin TG rats of 7 weeks old. Meanwhile, glycogen content in the liver tissues was decreased in regucalcin TG rats of 7 weeks old, suggesting that hepatic glycogen synthesis is suppressed in regucalcin TG rats [68]. The expression of regucalcin in the liver tissues was enhanced in regucalcin TG rats [44]. Regucalcin has been shown to have a suppressive effect on glycogen phosphorylase *a* activity in the glycogen particulate prepared from rat liver [89]. It is speculated that regucalcin suppresses glycogen synthesis in the liver and that the protein stimulates glycolysis in regucalcin TG rats. As the result, lipid synthesis may be stimulated in the liver tissues of regucalcin TG rats. Overexpression of regucalcin has been shown to enhance glucose utilization and lipid production in the cloned rat hepatoma H4-II-E cells *in vitro* [69]. In addition, regucalcin has been found to involve in insulin resistance in H4-II-E cells *in vitro* [69 - 71].

The expression of leptin mRNA in the adipose and liver tissues was found to decrease in regucalcin TG rats of 50 weeks old [68]. Adiponectin mRNA levels were not changed in the adipose tissues of regucalcin TG rats of 50 weeks old, while its expression was decreased in the liver tissues [68]. Leptin and adiponectin are adipokines that are involved in lipid metabolism [72, 73]. Leptin or adiponectin regulates hyperlipidemia [74, 75]. These decreases may be partly involved in the hyperlipidemia which is induced in regucalcin TG rats.

As mentioned above, hyperlipidemia that is induced in regucalcin TG rat may be resulted from the increase in hepatic lipid production and the suppression of leptin or adiponectin mRNA expression in the adipose and liver tissues in the TG rats as summarized in Figure 2.

OVEREXPRESSION OF REGUCALCIN ENHANCES HEPATIC LIPID PRODUCTION *IN VITRO*

Hyperlipidemia was induced in regucalcin TG rats [52], suggesting that lipid components are produced in the liver cells. To determine this, whether overexpression of regucalcin stimulates the production of lipids was examined using the cloned rat hepatoma H4-II-E cells *in vitro* [69]. Interestingly, insulin resistance was found to induce in the regucalcin-overexpressed rat hepatoma H4-II-E cells.

Overexpression of Regucalcin Enhances Glucose Utilization and Lipid Production in Hepatoma Cells *In Vitro*

The cloned rat hepatoma cells (wild type) and stable regucalcin/pCXN2-transfected cells (transfectant) were cultured for 72 hours in a medium containing 10% FBS to obtain subconfluent monolayers. Cells with subconfluency were cultured for 24 or 72 hours in a medium containing either vehicle or insulin (10^{-8} or 10^{-7} M) with or without supplementation of glucose (10, 25, or 50 mg/ml of medium) in the absence of insulin.

The productions of triglyceride and free fatty acid were significantly increased in the transfectants that were cultured without insulin and glucose supplementation as compared with that of wild-type cells [69]. The supplementation of glucose caused a remarkable increase in medium glucose consumption, triglyceride, and free fatty acid productions in the transfectants cultured without insulin [69]. These findings suggest that overexpression of regucalcin stimulates productions of triglyceride and free fatty acid in the H4-II-E cells. It is speculated that overexpression of regucalcin stimulates glucose utilization in the H4-II-E cells, and that it promotes the production of triglyceride and free fatty acid which are linked to glucose metabolism in the cells *in vitro* [69].

Insulin had a stimulatory effect on the consumption of medium glucose and the productions of triglyceride and free fatty acid in H4-II-E cells (wild type) cultured with or without the supplementation of glucose [69]. These increases were found to prevent in the transfectants. Overexpression of regucalcin may have a suppressive effect on insulin-stimulated glucose consumption in the H4-II-E cells, indicating insulin resistance.

The gene expression of acetyl-CoA carboxylase, HMG-CoA reductase, glucokinase and pyruvate kinase, which are rate-limiting enzymes related to glucose and lipid metabolism, in the H4-II-E cells (wild type) and the transfectans was not changed after culture with or without glucose supplementation in the presence of insulin [69]. The expression of glucose transporter 2 (GLUT 2) mRNA was found to increase in the transfectants as compared with that of wild-type cells cultured without insulin and glucose supplementation [69]. Overexpression of regucalcin did not have a stimulatory effect on the gene expression of enzymes, which are related to glucose and lipid metabolisms, in the H4-II-E cells. It is possible, however, that endogenous regucalcin has an activating effect on various enzyme activities, which are related to glucose and lipid metabolism, in H4-II-E cells. In addition, it is assumed that regucalcin has a regulatory role in signal transduction of insulin in H4-II-E cells. Regucalcin has been shown to have suppressive effects on protein tyrosine kinase and insulin action in liver cells [76].

Other studies show that insulin resistance may be modeled in H4-II-E liver cells in tissue culture with the use of the cytokine TNF-α and insulin [71]. From the proteome analysis of H4-II-E cells exposed to insulin and TNF-α, it has been proposed that regucalcin has a role associated with proteins that are involved in insulin resistance [77]. However, whether regucalcin can regulate the effect of insulin in liver cells was unknown. This was confirmed by our observations that overexpression of regucalcin induces insulin resistance in the H4-II-E cells [71, 77].

Both insulin resistance and increased oxidative stress in the liver are associated with the pathogenesis of nonalcoholic fatty liver disease. The pathogenesis of nonalcoholic fatty liver disease in patients has been reported to induce a decrease in hepatic regucalcin (RGN/SMP30) with a fibrosis stage-dependent manner and an increase in serum LDL, although it is not known whether decreased hepatic regucalcin is a result or a cause of cirrhosis [78]. The impairment of the early phase of insulin secretion in islets due to dysfunction of the distal portion of the secretion pathway has also been reported to underlie glucose intolerance in regucalcin knockout mice and decreased regucalcin may contribute to the worsening of glucose tolerance that occurs in normal aging [79]. These observations may be related to our early findings that overexpression of regucalcin enhances utilization of glucose and production of lipid components in the liver H4-II-E cells *in vitro* [69]. Presumably, regucalcin plays a part in the regulation of glucose and lipid metabolism.

Overexpression of Regucalcin Suppresses the Gene Expression of Insulin Signaling-Related Proteins in Hepatoma Cells *In Vitro*

As mentioned above, overexpression of regucalcin was shown to enhance glucose utilization and lipid production in the cloned rat hepatoma H4-II-E cells *in vitro*, and it induces insulin resistance [70]. The effect of regucalcin on the gene expression of insulin signaling-related proteins was examined using RT-PCR with specific primers in the cloned rat hepatoma cells overexpressing regucalcin *in vitro* [70].

The hepatoma cells (wild type) and stable regucalcin/pCXN2-transfected cells (transfectants) were cultured for 72 hours in a medium containing 10% FBS to obtain subconfluent monolayers. Cells with subconfluency were cultured for 24, 48, or 72 hours in a medium containing either vehicle or insulin (10^{-9} – 10^{-7} M) with or without supplementation of glucose (10, 25, or 50 mg/ml of medium). The expression of rat insulin receptor (Insr), phosphatidylinositol 3-kinase (PI3K), glucose transporter 2 (GLUT 2), or G3PDH mRNAs was determined. Overexpression of regucalcin was found to enhance GLUT 2 mRNA in the H4-II-E cells, although it failed to have a significant effect on the expression of Insr, PI3K, or G3PDH mRNAs in the cells cultured in the absence of insulin [70]. The expression of GLUT 2 mRNA in the transfectants may be partly involved in the enhancement of glucose utilization in the H4-II-E cells overexpressing regucalcin [70].

Overexpression of regucalcin, moreover, was found to have a suppressive effect on the expression of Insr and PI3K mRNAs that are enhanced after culture with insulin, glucose supplementation, or insulin plus glucose addition [70]. These observations suggest that overexpression of regucalcin suppresses gene expression of insulin signaling-related proteins, indicating that overexpression of regucalcin induces insulin resistance in the cloned rat hepatoma H4-II-E cells *in vitro*.

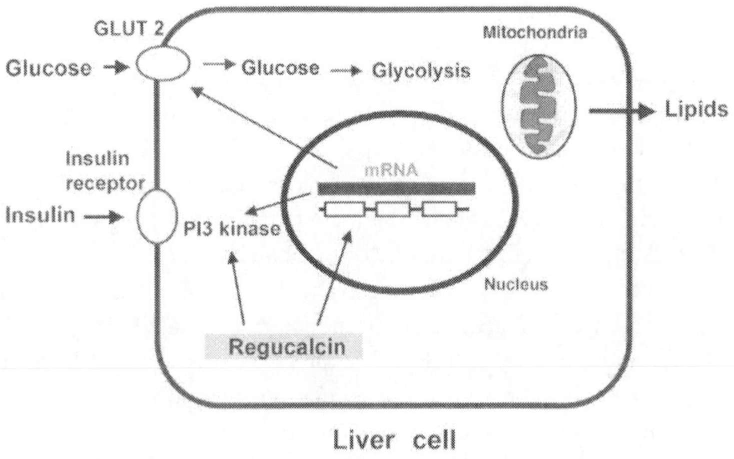

Figure 3. Overexpression of regucalcin stimulates glucose utilization and lipid production in the cloned rat hepatoma H4-II-E cells. Regucalcin increases GLUT 2 mRNA expression to enhance glucose utilization in the cells. Regucalcin suppresses the gene expression of Insr or PI3K that is an insulin signaling-related protein, which is enhanced after culture with insulin and/or glucose supplementation. Overexpression of regucalcin induces insulin resistance.

The mRNA expression of acetyl-CoA carboxylase, HMG-CoA reductase, glucokinase, and pyruvate kinase, which are related to glucose and lipid metabolism in liver cells, in the cloned rat hepatoma H4-II-E cells was not changed after culture with or without glucose supplementation in the presence of insulin [70]. The suppressive effect of regucalcin on Insr and PI3K mRNA expression may be important in inducing insulin resistance in the cloned rat hepatoma H4-II-E cells overexpressing regucalcin.

In conclusion, regucalcin TG rat induced a remarkable increase in lipid components in the serum [52]. The decrease in lipid component and glycogen in the liver tissues was found in regucalcin TG rats *in vivo* [68]. Moreover, overexpression of regucalcin enhanced glucose utilization and lipid production in the cloned rat hepatoma H4-II-E cell model *in vitro* and it suppressed the effect of insulin in the cells. Overexpression of regucalcin in the H4-II-E cells suppressed the gene expression of Insr and PI3K mRNAs that are related to insulin siganaling in liver cells, inducing insulin resistance [70]. The regulatory effect of regucalcin in the liver tissues, which is related to insulin signaling, is may be important in hyperlipidemia induced in regucalcin TG rats as summarized in Figure 3.

Thus, the disorder of lipid metabolism has been demonstrated to induce in regucalcin TG rats, indicating its usefulness as an animal model of hyperlipidemia.

CONCLUSION

Regucalcin has been demonstrated to play a multifunctional role as a regulatory protein in Ca^{2+} signaling and other signaling processes in many cell types [reviewed in references 15-20]; maintaining intracellular Ca^{2+} homeostasis; inhibiting protein kinases and protein phosphatases; suppressing protein synthesis and DNA and RNA synthesis; inhibiting cell proliferation; and suppressing apoptotic cell death. From these findings, regucalcin has been proposed to play a pivotal role in maintaining cell homeostasis and function as the regulatory protein of intracellular signaling system [20].

Overexpression of regucalcin has been found to induce osteoporosis and hyperlipidemia in the TG rats with weanling and aging, indicating that regucalcin plays a physiological role in the regulation of bone homeostasis and lipid metabolism. The exploration of the mechanism by which regucalcin TG rat induces bone loss and hyperlipidemia proposes the information that regucalcin is a novel protein which contributes in the regulation of bone and lipid metabolisms, suggesting a physiological role of regucalcin *in vivo*.

Moreover, hyperphosphatemia, hypercalcemia, and hyperalbuminemia associated with osteoporosis and hyperlipidemia has been found to induce in regucalcin TG rats. Especially, the suppression of leptin and adiponectin mRNAs expression was induced in regucalcin TG rats *in vivo*. These hormones are involved in the regulation of lipid metabolism. Regucalcin may play an important role in the regulation of lipid metabolism *in vivo*.

Presumably, regucalcin plays a role in the regulation of various functions in the living body. Regucalcin TG rat may be a good animal model in the exploration of mechanism of metabolic disorders and the development of their therapeutic tools.

REFERENCES

[1] Cheung, W.Y. (1980). Calmodulin plays a pivotal role in cellular regulation. *Science*, 202: 19-27.

[2] Nishizuka, Y. (1986). Studies and perspectives of protein kinase C. *Science*, 233: 305-312.

[3] Kraus-Friedman, N., Feng, L. (1996). The role of intracellular Ca^{2+} in the regulation of gluconeogenesis. *Metabolism*, 42: 389-403.

[4] Yamaguchi, M., Yamamoto, T. (1978). Purification of calcium binding substance from soluble fraction of normal rat liver. *Chem. Pharm. Bull.,* 26: 1915-1918.

[5] Yamaguchi, M., Sugii, K. (1981). Properties of calcium-binding protein isolated from the soluble fraction of normal rat liver. *Chem. Pharm. Bull.,* 29: 567-570.

[6] Yamaguchi, M. (1988). Physicochemical properties of calcium-binding protein isolated from rat liver cytosol: Ca^{2+}-induced conformational changes. *Chem. Pharm. Bull.,* 36: 286-290.

[7] Shimokawa, N., Yamaguchi, M. (1993). Molecular cloning and sequencing of the cDNA coding for a calcium-binding protein regucalcin from rat liver. *FEBS Lett*., 327: 251-255.

[8] Yamaguchi, M., Yoshida, H. (1985). Regulatory effect of calcium-binding protein isolated from rat liver cytosol on activation of fructose 1,6-diphosphatase by Ca^{2+}-calmodulin. *Chem. Pharm. Bull.,* 33: 4489-4493.

[9] Yamaguchi, M., Shibano, H. (1987). Calcium-binding protein isolated from rat liver cytosol reverses activation of pyruvate kinase by Ca^{2+}. *Chem. Pharm. Bull.,* 35: 2025-2029.

[10] Yamaguchi, M., Shibano, H. (1987). Effect of calcium-binding protein on the activation of phosphorylase *a* in rat hepatic particulate glycogen by Ca^{2+}. *Chem. Pharm. Bull.,* 35: 2581-2584.

[11] Yamaguchi, M., Shibano, H. (1987). Reversible effect of calcium-binding protein on the Ca^{2+}-induced activation of succinate dehydrogenase in rat liver mitochondria. *Chem. Pharm. Bull.,* 35: 3766-3770.

[12] Yamaguchi, M., Mori, S. (1988). Effect of Ca^{2+} and Zn^{2+} on 5'-nucleotidase activity in rat liver plasma membranes: Hepatic calcium-binding protein (regucalcin) reverses the Ca^{2+} effect. *Chem. Pharm. Bull.,* 36: 321-325.

[13] Yamaguchi, M. (1992). A novel Ca^{2+}-binding protein regucalcin and calcium inhibition. Regulatory role in liver cell function. In: Kohama K (ed) *Calcium inhibition*, Japan Sci. Soc. Press, Tokyo and CRC Press, Boca Raton pp.19-41.

[14] Yamaguchi, M. (1998). Role of calcium-binding protein regucalcin in regenerating rat liver. J. Gastroen. Hepatol., 13 (Suppl.): S106-S112.

[15] Yamaguchi, M. (2000). Role of regucalcin in calcium signaling. *Life Sci.,* 66: 1769-1780.

[16] Yamaguchi, M. (2002). The role of regucalcin in nuclear regulation of regenerating liver. *Biochem. Biophys. Res. Commun.,* 276: 1-6, 2000.

[17] Yamaguchi, M. (2002). Impact of aging on calcium channels and pumps. In: Mattson MP (ed) *Calcium homeostasis and signaling in aging*. Elsevier, Amsterdam pp. 47-65.

[18] Yamaguchi, M. (2005). Role of regucalcin in maintaining cell homeostasis and function. *Int. J. Mol. Med.,* 15: 372-389.

[19] Yamaguchi M (2011). The transcriptional regulation of regucalcin gene expression. *Mol. Cell. Biochem*., 346:147-171.

[20] Yamaguchi, M. (2011). Regucalcin and cell regulation: role as a suppressor protein in signal transduction. *Mol. Cell. Biochem*., 353:101-137.

[21] Misawa, H., Yamaguchi, M. (2000). The gene of Ca^{2+}-binding protein regucalcin is highly conserved in vertebrate species. *Int. J. Mol. Med.,* 6: 191-196.

[22] Nikapitiya, C., De Zoysa, M., Kang, H.S., Oh, C., Whang, I., Lee, J. (2008). Molecular characterization and expression analysis of regucalcin in disk abalone (Haliotis discus discus): intramuscular calcium administration stimulates the regucalcin mRNA expression. *Comp. Biochem. Physiol. B Biochem. Mol. Biol.,* 150: 117-124.

[23] Shimokawa, N., Matsuda, Y., Yamaguchi, M. (1995). Genomic cloning and chromosomal assignment of rat regucalcin gene. *Mol. Cell. Biochem.*, 151:157-163.

[24] Thiselton, D.L., McDowall, J., Brandau, O., Ramser, J., d'Esposito, F., Bhattacharga, S.S., Ross, M.T., Hardcastle, A.J., Meindl, A. (2002). An integrated, functionally annotated gene map of the DXS8026-ELK1 internal on human Xp11.3-Xp11.23: Potential hotspot for neurogenetic disorders. *Genomics*, 79: 560-572.

[25] Yamaguchi, M., Makino, R., Shimokawa, N. (1996). The 5'end seguences and exon organization in rat regucalcin gene. *Mol. Cell. Biochem.,* 165: 145-150.

[26] Murata, T., Yamaguchi, M. (1998). Ca^{2+} administration stimulates the binding of AP-1 factor to the 5'-flanking region of the rat gene for the Ca^{2+}-binding protein regucalcin. *Biochem. J.,* 329: 157-163.

[27] Miasawa, H., Yamaguchi, M. (2000). Involvement of hepatic nuclear factor I binding motif in transcriptional regulation of Ca^{2+}-binding protein regucalcin gene. *Biochem. Biophys. Res. Commun.,* 269: 270-278.

[28] Misawa, H., Yamaguchi, M. (2002). Indentification of transcription factor in the promoter region of rat regucalcin gene: Binding of nuclear factor I-A1 to TTGGC motif. *J. Cell. Biochem.,* 84: 795-802.

[29] Misawa, H., Yamaguchi, M. (2001). Molecular cloning and sequencing of the cDNA coding for a novel regucalcin gene promoter region-related protein in rat, mouse and human liver. *Int. J. Mol. Med.,* 8: 513-520.

[30] Sawada, N., Yamaguchi, M. (2005). A novel regucalcin gene promoter region-related protein: Comparison of nucleotide and amino acid sequnces in vertebrates species. *Int. J. Mol. Med.,* 15: 97-104.

[31] Yamaguchi, M. (2009). Novel protein RGPR-p117: its role as the regucalcin gene transcription factor. *Mol. Cell. Biochem.,* 327: 53-63.

[32] Nejak-Bowen, K.N., Zeng, G., Tan, X., Cieply, B., Monga, S.P. (2009). Beta-catenin regulates vitamin C biosynthesis and cell survival in murine liver. *J. Biol. Chem.,* 284: 28115-28127.

[33] Murata, T., Yamaguchi, M. (1999). Promoter characterization of the rat gene for Ca^{2+}-binding protein regucalcin. Transcriptional regulation by signaling factors. *J. Biol. Chem.,* 274: 1277-1285.

[34] Misawa, H., Yamaguchi, M. (2000). Intracellular signaling factors-enhanced hepatic nuclear protein binding to TTGGC sequence in the rat regucalcin gene promoter: involvement of protein phosphorylation. *Biochem. Biophys. Res. Commun.,* 279: 275-281.

[35] Misawa, H., Yamaguchi, M. (2002). Gene expression for a novel protein RGPR-p117 in various species: The stimulation by intracellular signaling factors. *J. Cell. Biochem.,* 87: 188-193.

[36] Sawada, N., Yamaguchi, M. (2006). Overexpression of RGPR-p117 enhances regucalcin gene promoter activity in cloned normal rat kidney proximal tubular epithelial cells: Involvement of TTGGC motif. *J. Cell. Biochem.,* 99: 589-597.

[37] Yamaguchi, M., Nakajima, M. (1999). Involvement of intracellular signaling factors in the serum-enhanced Ca^{2+}-binding protein regucalcin mRNA expression in the cloned rat hepatoma cells (H4-II-E). *J. Cell. Biochem.,* 74: 81-89.

[38] Nakagawa, T., Yamaguchi, M. (2005). Hormonal regulation on regucalcin mRNA expression in cloned normal rat kidney proximal tubular epithelial NRK52E cells. *J. Cell. Biochem.,* 95: 589-597.

[39] Yamaguchi, M., Isogai, M. (1993). Tissue concentration of calcium-binding protein regucalcin in rats by enzyme-linked immunoadsorbent assay. *Mol. Cell. Biochem.,* 122: 65-68.

[40] Shimokawa, N., Isogai, M., Yamaguchi, M. (1995). Specific species and tissue differences for the gene expression of calcium-binding protein regucalcin. *Mol. Cell. Biochem.,* 143: 67-71.

[41] Shimokawa, N., Yamaguchi, M. (1992). Calcium administration stimulates the expression of calcium-binding protein regucalcin mRNA in rat liver. *FEBS Lett.*, 305: 151-154.

[42] Yamaguchi, M., Hamano, T., Misawa., H. (2000). Expression of Ca^{2+}-binding protein regucalcin in rat brain neurons: inhibitory effect on protein phosphatase activity. *Brain. Res. Bull.,* 52: 343-348.

[43] Yamaguchi, M., Nakajima, R. (2002). Role of regucalcin as an activator of sarcoplasmic reticulum Ca^{2+}-ATPase activity in rat heart muscle. *J. Cell. Biochem.,* 86: 184-193.

[44] Yamaguchi, M., Misawa, H., Uchiyama, S., Morooka, Y., Tsurusaki, Y. (2002). Role of endogenous regucalcin in bone metabolism: Bone loss is induced in regucalcin transgenic rats. *Int. J. Mol. Med.*, 10: 377-383.

[45] Maia, C.J., Santos, C.R., Schmitt, F., Socorro, S. (2008). Regucalcin is expressed in rat mammary gland and prostate and down-regulated by 17beta-estradiol. *Mol. Cell. Biochem.*, 311: 81-86.

[46] Maia, C.J., Santos, C.R., Schmitt, F., Socorro, S. (2009). Regucalcin is under-expressed in human breast and prostate cancers: Effect of sex steroid hormones. *J. Cell. Biochem.,* 107: 667-676.

[47] Murata, T., Shinya, N., Yamaguchi, M. (1997). Expression of calcium-binding protein regucalcin mRNA in the cloned human hepatoma cells (HepG2): stimulation by insulin. *Mol. Cell. Biochem.,* 175: 163-168.

[48] Yamaguchi, M., Oishi, K. (1995). 17β-Estradiol stimulates the expression of hepatic calcium-binding protein regucalcin mRNA in rats. *Mol. Cell. Biochem.,* 143: 137-141.

[49] Kondo, Y., Inai, Y., Sato, Y., Handa, S., Kubo, S., Shimokado, K., Goto, S., Nishikimi, M., Maruyama, N., Ishigami, A. (2006). Senescence marker protein 30 functions as gluconolactonase in L-ascorbic acid biosynthesis, and its knockout mice are prone to scurvy. *Proc. Natl. Acad. Sci., USA* 103: 5723-5728.

[50] Yamaguchi, M., Morooka, Y., Misawa, H., Tsurusaki, Y., Nakajima, R. (2002) Role of endogenous regucalcin in transgenic rats: Suppression of kidney cortex cytosolic protein phosphatase activity and enhancement of heart muscle microsomal Ca^{2+}-ATPase activity. *J. Cell. Biochem.,* 86: 520-529.

[51] Morooka, Y., Yamaguchi, M. (2001). Suppressive role of endogenous regucalcin in the regulation of protein phosphatase activity in rat renal cortex cytosol. *J. Cell. Biochem.,* 81: 639-646.

[52] Yamaguchi, M., Igarashi, A., Uchiyama, S., Sawada, N. (2004). Hyperlipidemia is induced in regucalcin transgenic rats with increasing age. *Int. J. Mol. Med.,* 14: 647-651.

[53] Tsurusaki, Y., Yamaguchi, M. (2003). Role of endogenous regucalcin in transgenic rats: Suppression of protein tyrosine phosphatase and ribonucleic acid synthesis activities in liver nucleus. *Int. J. Mol. Med.,* 12: 207-211.

[54] Ma, Z.J., Yamaguchi, M. (2003). Regulatory effect of regucalcin on nitric oxide synthase activity in rat kidney cortex cytosol: Role of endogenous regucalcin in transgenic rats. *Int. J. Mol. Med.*, 12: 201-206.

[55] Ma, Z.J., Yamaguchi, M. (2002). Suppressive role of endogenous regucalcin in the regulation of nitric oxide synthase activity in heart muscle cytosol of normal and regucalcin transgenic rats. *Int. J. Mol. Med.,* 10: 761-766.

[56] Tobisawa, M., Yamaguchi, M. (2003). Role of endogenous regucalcin in brain function: suppression of cytosolic nitric oxide synthase and nuclear protein tyrosine phosphatase activities in brain tissue of transgenic rats. *Int. J. Mol. Med.,* 12: 581-585.

[57] Weinstock, P.H., Bisgaier, C.L., Aalto-Setala, K., Radner, H., Ramarkrishnan, R., Levak-Frank, S., Essenbury, A.D., Zechner, R., Breslow, J.L. (1995). Severe hypertriglyceridemia, reduced high density lipoprotein, and neonatal death in lipoprotein lipase in knockout mice. Mild hypertriglyceridemia with impaired very low density lipoprotein clearance in heterozygotes. *J. Clin. Invest.,* 96: 2555-2568.

[58] Lichtman, A.H., Clinton, S.K., Iiyama, K., Connelly, P.W., Libby, P., Cybulsky, M.I. (1999). Hyperlipidemia and atherosclerotic lesion development in LDL receptor-deficient mice fed defined semipurified diets with and without cholate. *Arterioscler. Thromb. Vasc. Biol.,* 19: 1938-1944.

[59] Jong, M.C., Havekes, L.M. (2000). Insights into apolipoprotein C metabolism from transgenic and gene-targeted mice. *Int. J. Tissue React.,* 22: 59-66.

[60] Koopmans, S.J., Jong, M.C., Que, I., Dahlmans, V.E., Pijl, H., Radder, J.K., Frolich, M., Havekes, L.M. (2001). Hyperlipidemia is associated with increased insulin-mediated glucose metabolism, reduced fatty acid metabolism and normal blood pressure in transgenic mice overexpressing human apolipoprotein C1. *Diabetologia*, 44: 437-443.

[61] Yagyu, H., Lutz, E.P., Kako, Y., Marks, S., Hu, Y., Choi, S.Y., Bensadoun, A., Goldberg, I.J. (2002). very low density lipoprotein (VLDL) receptor-deficient mice have reduced lipoprotein lipase activity mass with VLDL receptor deficincy. *J. Biol. Chem.,* 277: 10037-10043.

[62] Chen, J.Y., Levy-Wilson, B., Goodart, S., Cooper, A.D. (2002). Mice expressing the human CYP7A1 gene in the mouse CYP7A1 knock-out background lock induction of CYP7A1 expression by cholesterol feeding and have increased hypercholesterolemia when fed a high fat diet. *J. Biol. Chem.,* 277: 42588-42595.

[63] Fazio, S., Linton, M.F. (2001). Mouse models of hyperlipidemia and atheroscerosis. *Front. Biosci.,* 6: D515-D525.

[64] Ono, H., Shibano, H., Katagiri, H., Yahagi, N., Sakoda, H., Onishi, Y., Anai, M., Ogihara, T., Fujishiro, M., Viana, A.Y., Fukushima, Y., Abe, M., Shojima, N., Kikuchi, M., Yamada, N., Oka, Y., Asano, T. (2003). Hepatic Akt activation induces marked hypoglycemia, hepatomegaly, and hypertriglyceridemia with sterol regulatory element binding protein involvement. *Diabetes*, 52: 2905-2913.

[65] Yamaguchi, M., Shibano, H. (1987). Effect of calcium-binding protein on the activation of phosphorylase *a* in rat hepatic particulate glycogen by Ca^{2+}. *Chem. Pharm. Bull.,* 35: 2581-2584.

[66] Yamaguchi, M., Shibano, H. (1987). Calcium-binding protein isolated from rat liver cytosol reverses activation of pyruvate kinase by Ca^{2+}. *Chem. Pharm. Bul.,l* 35: 2025-2029.

[67] Yamaguchi, M., Yoshida, H. (1985). Regulatory effect of calcium-binding protein isolated from rat liver cytosol on activation of fructose 1,6- diphosphatase by Ca^{2+}-calmodulin. *Chem. Pharm. Bull.,* 33: 4489-4493.

[68] Yamaguchi, M., Nakagawa, T. (2007). Change in lipid components in the adipose and liver tissues of regucalcin transgenic rats with increasing age: Suppression of leptin and adiponectin gene expression. *Int. J. Mol. Med.,* 20: 323-328.

[69] Nakashima, C., Yamaguchi, M. (2006). Overexpression of regucalcin enhances glucose utilization and lipid production in cloned rat hepatoma H4-II-E cells: Involvement of insulin resistance. *J. Cell. Biochem.,* 99: 1582-1592.

[70] Nakashima, C., Yamaguchi, M. (2007). Overexpression of regucalcin suppresses gene expression of insulin siganaling-related proteins in cloned rat hepatoma H4-II-E cells: Involvement of insulin resistance. *Int. J. Mol. Med.,* 20: 709-716.

[71] Soloman, S.S., Buss, N., Shull, J., Monnier, S., Majumdar, G., Wu, J., Gerling, I.C. (2005). Proteome of H4-II-E (liver) cells exposed to insulin and tumor necrosis factor-α: Analysis of proteins involved in insulin resistance. *J. Lab. Clin. Med.,* 145: 275-283.

[72] Roni, T., Lupattelli, G., Mannarino, E. (2006). The endocrine function of adipose tissue: an uptake. *Clin. Endocrinol.,* 64: 255-365.

[73] Dyck, D.J., Heigenhauser, G.J., Bruce, C.R. (2006). The role of adipokines as regulators of skeletal muscle fatty acid metabolism and insulin sensitivity. *Acta Physio.,l* 186: 5-16.

[74] Guerre-Millo, M. (2002). Adipose tissue hormones. *J. Endocrinol., Invest.*, 25: 855-861.

[75] Havel, P.J. (2004). Update on adipocyte hormones: regulation of energy balance and carbohydrate/lipid metabolism. *Diabetes* 53: 5143-5151.

[76] Fukaya, Y., Yamaguchi, M. (2005). Overexpression of regucalcin suppresses cell death and apoptosis in cloned rat hepatoma H4-II-E cells induced by insulin or insulin-like growth factor-I. *J. Cell. Biochem.,* 96: 145-154.

[77] Solomon, S.S., Mishra, S.K., Palazzolo, M.R., Postlethwaite, A.E., Seyer, J.M. (1997). Identification of specific sites in the TNF-α molecule promoting insulin resistance in H4-II-E liver cells. *J. Lab. Clin. Med.,* 130: 139-146.

[78] Park, H., Ishigami, A., Shima, T., Mizuno, M., Maruyama, N., Yamaguchi, K., Mitsuyoshi, H., Minami, M., Yasui, K., Itoh, Y., Yoshikawa, T., Fukui, M., Hasegawa, G., Nakamura, N., Ohta, M., Obayashi, H., Okanoue, T. (2009). Hepatic senescence marker protein-30 is involved in the progression of nonalcoholic fatty liver disease. *J. Gastroenterol.*, 45:426-434.

[79] Hasegawa, G., Yamasaki, M., Kadono, M., Tanaka, M., Asano, M., Senmaru, T., Kondo, Y., Fukui, M., Obayashi, H., Maruyama, N., Nakamura, N., Ishigami, A. (2010). Senescence marker protein-30/gluconolactonase deletion worsens glucose tolerance through impairment of acute insulin secretion. *Endocrinology* 151: 529-536.

In: Adiponectin: Production, Regulation and Roles in Disease ISBN: 978-1-61942-165-3
Editor: Masayoshi Yamaguchi

Chapter VIII

EFFECTS OF LOSARTAN/HYDROCHLOROTHIAZIDS ON ADIPONECTIN AND ANGIOPATHY-RELATED MARKERS IN HYPERTENSIVE PATIENTS WITH TYPE 2 DIABETES MELLITUS

Shosaku Nomura[a1], Norihito Inami[b], Seitarou Omoto[b], Takayuki Shimazu[b], Takayuki Kajiura[b], Kohichi Yamada[b], Ryotaro Ozasa[a], Shinya Fujita[a], Takashi Yokoi[a] and Akira Shouzu[b]

[a]First Department of Internal Medicine, Kansai Medical University, Japan
[b]Second Department of Internal Medicine, Kansai Medical University, Japan

ABSTRACT

Angiotensin II receptor blockade has been shown to have a beneficial effect on the angiopathies of hypertension and hyperglycemia in patients with type 2 diabetes. On the other hand, blood pressure is difficult to control with monotherapy, especially in patients with diabetes, highlighting the requirement of drug contribution such as losartan/hydrochlorothiazide (LST/HCTZ). However, the effect of LST/HCTZ on the angiopathy-related markers in hypertension and type 2 diabetes is poorly understood. We investigated the effects of LST/HCTZ on these markers in 82 hypertensive patients with and without type 2 diabetes mellitus. Levels of the angiopathy-related markers (platelet-derived microparticles: 11.6 ± 2.7 vs. 17.1 ± 5.6 IU/ml, $p < 0.01$; RANTES: 73.1 ± 12.2 vs. 88.8 ± 20.4 ng/ml, $p < 0.01$; monocyte chemotactic peptide-1: 430 ± 104 vs. 492 ± 120 pg/ml, $p < 0.05$; and soluble E-selectin: 48 ± 19 vs. 62 ± 22 ng/ml, $p < 0.01$) were significantly increased in hypertensive patients with type 2 diabetes compared to without type 2 diabetes. In addition, the concentrations of adiponectin were significantly decreased in patients with type 2 diabetes (4.71 ± 0.85 vs. 3.19 ± 1.14 μg/ml, $p < 0.01$). Regardless of the presence of diabetic complications, both systolic and diastolic blood pressure significantly decreased after LST/HCTZ administration (LST/HCTZ 50 mg/day for 12 weeks). The angiopathy-related markers were decreased significantly in patients

[1] E-mail address: shosaku-n@mbp.ocn.ne.jp

with type 2 diabetes after LST/HCTZ treatment, but not in non-type 2 diabetic patients. In addition, LST/HCTZ alleviated hypoadiponectinemia in hypertensive patients with diabetes (before vs. after: 3.19 ± 1.14 vs. 4.28 ± 0.93 mg/ml, $p < 0.01$) but did not increase adiponectin levels in the non-diabetic hypertensive group, for which the average adiponectin level was normal prior to treatment. These results suggest LST/HCTZ may be beneficial as an adiponectin-dependent anti-atherosclerotic therapy in patients with type 2 diabetes in addition to its anti-hypertensive action.

Keywords: Angiopathy-related marker, adiponectin, hypertension, type 2 diabete mellitus, LST/HCTZ

INTRODUCTION

The UK Prospective Diabetes Study report [1] states that microvascular damage, and all diabetes mellitus-related events, including death, stroke, and nephropathy, are reduced when blood pressure is closely controlled. Diabetes mellitus leads to the development of premature atherosclerosis [2,3], and the atherosclerosis associated with diabetes accelerates diabetic macroangiopathy [4,5]. Patients with diabetes mellitus are at increased risk of hyperlipidemia and hypertension [6]. It has been well established that blood pressure control as well as glycemic control is important in the management of these individuals [1].

Activated platelets may be a potential cause of diabetic vascular complications. In addition, we previously reported that high levels of platelet-derived microparticle (PDMP) might be a marker for vasculopathies in diabetic patients [7-9]. Monocyte chemotactic peptide 1 (MCP-1) is a C-C chemokine for monocytes [10,11], and it has been reported that MCP-1 may contribute to glomerular monocyte infiltration in diabetic nephropathy [12]. Soluble (s)E-selectin is a type of endothelial cell dysfunction markers that are significantly increased in diabetic vascular complications [13].

Adiponectin, the most abundant adipose-specific protein, is exclusively expressed in and secreted from adipose tissue [14]. Plasma adiponectin concentrations are decreased in obese individuals [14,15] with type 2 diabetes [16] and are closely related to whole-body insulin sensitivity [17]. This protein is abundant in the circulation [15] and suppresses the attachment of monocytes to endothelial cells [16]. Adiponectin also stimulates nitric oxide production in vascular endothelial cells, which ameliorates endothelial dysfunction [19,20]. These observations suggest anti-atherogenic properties for adiponectin, and that hypoadiponectinemia might be associated with a higher incidence of vascular disease in diabetic subjects.

Angiotensin II plays a central role in the pathogenesis of hypertension-induced end-organ damage [21,22]. In addition, it is suggested that angiotensin II could also induce endothelial dysfunction through the production of oxidative stress [23] and by enhancing leukocyte adhesion to endothelial cells [24]. Furthermore, there is accumulating evidence of the superiority of angiotensin II receptor blockers (ARBs) on reducing microalbuminuria or proteinuria [25,26], which is a risk factor for renal and cardiovascular diseases [27,28]. Therefore, ARB is recommended as one of the first-line antihypertensive agent in patients with diabetes mellitus [29]. However, to our knowledge, few investigators have studied the relationship between losartan/hydrochlorothiazide (LST/HCTZ), adiponectin and angiopathy-related markers. In the present study, we investigated the effects of treatment with

LST/HCTZ on angiopathy-related markers and adiponectin in patients with type 2 diabetes mellitus, since LST/HCTZ may prevent the progression of angiopathy in hypertensive patients with diabetes.

MATERIALS AND METHODS

Patients

The study group included 25 controls and 82 hypertensive patients. Controls were recruited from our hospital staff and other sources. Hypertensive patients were selected from patients admitted to our hospital for assessment of hypertension between September 2007 and April 2010. The study protocol was approved by the Institutional Review Board (IRB) of our institution and written informed consent was obtained from each subject prior to the start of the trial. Within the past 3 months, none of the study participants had experienced inflamematory, coronary artery, or cerebrovascular disease or had a clinically detectable renal (serum creatinine < 2.0 mg/dl), hepatic (normal serum transamilase), infectious (no fever and normal white blood cell), or malignant disease (normal ultrasound or computed tomography). Antithrombotic agents, including aspirin, were withheld, since these drugs might affect the present data. There were seventeen patients using a statin medication and eleven patients using an ACE inhibitor. However, the ACE inhibitor medication had been suspended for more than 2 weeks before LST/HCTZ treatment. thirty-nine of the hypertensive patients had type 2 diabetes and 43 did not. Ten patients were diagnosed with nephropathy. Diabetic nephropathy was diagnosed by the presence of macroalbuminurea (> 300 mg/day) in more than three consecutive urine samples. Treatment of diabetes was with sulfonylureas (12/43), α -glucosidase inhibitors (21/43), and insulin therapy (8/43). The diagnosis criteria for hypertension were a recumbent systolic blood pressure of > 150 mm Hg and a recumbent diastolic pressure of > 90 mm Hg on two or more occasions [31]. Type 2 diabetes was defined according to the American Diabetes Association Criteria [32].

Study Design

We administered LST/HCTZ (MSD Pharmaceutical, Tokyo, Japan), 50 mg/day, to the hypertensive patients for 12 weeks. There were no other changes in any patient's pharmacologic regimen during valsartan treatment. Clinical and biochemical data obtained before and after LST/HCTZ administration were compared.

Assessment of PDMP

An enzyme-linked immunosorbent assay (ELISA) kit for the detection of PDMP [7,9] was obtained from Jimro Co., Ltd. (Tokyo, Japan). In brief, a blood sample was collected from a peripheral vein into a vacutainer containing EDTA-ACD (Nipro Co. Ltd., Japan) with 21-gauge needle to minimize platelet activation. The sample was gently mixed by turning the tube upside down once or twice and then was left at room temperature for 2-3 hr, followed by centrifugation at 8,000 g for 5 min at room temperature. Immediately after centrifugation, we collected 200 μl of the upper layer of supernatant from a 2-ml sample to avoid contamination and stored each sample at –40°C until analysis. The results of ELISA performed under the current experimental conditions were reproducible. PDMP were measured twice and the mean value was calculated. The kit employed two monoclonal antibodies directed against platelet glycoproteins CD42b and CD42a (glycoprotein Ib and IX). One U/ml of PDMP was defined as 24,000 solubilized platelets/ml in this ELISA.

Measurement of Regulated on Activation Normally T-cell Expressed and Secreted (RANTES), MCP-1, sE-selectin and Adiponectin

Blood samples from patients and controls were collected into tubes with sodium citrate or tubes without anticoagulant. Blood samples were allowed to clot at room temperature for a minimum of 1 hour. Serum or citrated plasma was isolated by centrifugation for 20 minutes at 1,000 g (4°C) and stored at -30°C until analysis. The recombinant products and standard solutions provided with the commercial kits were used as positive controls in each assay. The specific ELISA kits for RANTES, MCP-1 and sE-selectin were obtained from BioSource International, Inc. (CA), and the Adiponectin ELISA kit from Otsuka Pharmaceuticals Co. Ltd (Tokyo, Japan). All kits were used according to the manufacturer's instructions.

Statistics

Data are presented as the mean ± SD. The significances among variables were determined by analysis of variance (ANOVA). Student's *t*- test was used for statistical comparisons, and *P*-values < 0.05 were considered significant.

RESULTS

Table 1 shows the clinical characteristics of the controls and hypertensive subjects. Systolic and diastolic blood pressure values prior to treatment were similar in both the diabetes-positive and negative subgroups of the hypertensive patients (Table 2). However, the angiopathy-related markers were higher in hypertensive patients with type 2 diabetes than in controls (Table 2, PDMP: 5.2 ± 0.6 vs. 17.1 ± 5.6 U/ml, p < 0.01; RANTES: 55.7 ± 3.8 vs. 88.4 ± 20.4 ng/ml, p < 0.01; MCP-1: 388 ± 87 vs. 492 ± 120 pg/ml, p < 0.05; sE-selectin: 42 ± 10 vs. 62 ± 22 ng/ml, p < 0.001). Adiponectin levels in hypertensive patients with diabetes prior to LST/HCTZ treatment were significantly lower than levels in controls (Table 2, Adiponectin: 5.93 ± 1.12 vs. 3.19 ± 1.14 μg/ml, p < 0.01). In addition, 10 diabetic patients had diabetic nephropathy complications. Although the angiopathy-related markers of diabetic patients with nephropathy were significantly increased compared to patients without such complications (nephropathy-positive vs. -negative, number of patients: 10 vs. 29, PDMP: 19.4 ± 4.9 vs. 15.6 ± 3.8 U/ml, p < 0.05; MCP-1: 523 ± 126 vs. 451 ± 130 pg/ml, p < 0.05;

RANTES: 95.6 ± 19.2 vs. 79.1 ± 19.5 ng/ml, $p < 0.05$; sE-selectin: 78 ± 25 vs. 44 ± 19 ng/ml, $p < 0.01$). There were no noticeable differences in the lowering effect of LST/HCTZ on angiopathy-related markers among the groups. In addition, we also could not find the differences among patients using sulfonylureas, a-glucosidase inhibitors, and insulin therapy. In this study, 20 patients were on a statin therapy. However, they were evenly distributed among the non-diabetic and diabetic groups, and values for blood pressure, and angiopathy-related markers did not differ significantly between patients using statins and those who were not.

Table 1. Demographic and Clinical Characteristics of Controls and Hypertensive Patients

	control (n=25)	Hypertensive (n=82)	
		DM(-)(n=43)	DM(+)(n=39)
Gender (male/female)	15/10	13/30	11/28
Age, y	53 ± 6	62 ± 6	66 ± 9
BMI, kg/m^2	23.2 ± 4.3	23.8 ± 3.9	24.6 ± 5.3
Smokers, n (%)	10 (40)	19 (44.2)	17 (43.6)
WBC, / μ L	5,170 ± 790	4,980 ± 977	6,650 ± 1,870
PLT, × 10^4/ μ L	26.8 ± 4.6	27.2 ± 5.5	26.8 ± 6.5
HbA1c, %	4.5 ± 2.0	5.4 ± 2.5	9.7 ± 2.6 ***
TC, mmol/L	5.34 ± 1.09	5.74 ± 1.13	6.22 ± 1.35 **
TG, mmol/L	1.92 ± 0.33	2.36 ± 0.44 *	2.37 ± 0.64 *
LDL, mmol/L	3.45 ± 1.02	3.55 ± 1.11	3.68 ± 1.19 **
HDL, mmol/L	1.29 ± 0.45	1.24 ± 0.47	1.15 ± 0.36 *

Control: non-diabetes Data are shown as mean ± SD.
BMI: body mass index; WBC: white blood cell count; PLT: platelet count ;
HbA1c (%): hemoglobin A1c TC: total cholesterol ; TG: triglycerides
LDL: low-density lipoprotein HDL: high-density lipoprotein
* $p < 0.05$ ** $p < 0.01$ *** $p < 0.001$; versus controls

Both diastolic and the systolic blood pressure decreased significantly after LST/HCTZ administration in both groups (Figure 1: non-diabetes, systolic blood pressure 190 ± 15 vs. 165 ± 14 mmHg, $p < 0.001$, diastolic blood pressure 95 ± 11 vs. 88 ± 10 mmHg, $p < 0.01$; Figure 1: diabetes, systolic blood pressure 189 ± 14 vs. 165 ± 12 mmHg, $p < 0.001$, diastolic blood pressure 99 ± 10 vs. 91 ± 7 mmHg, $p < 0.01$). Additionally, hypertensive patients with type 2 diabetes exhibited a significant increase in adiponectin levels after LST/HCTZ treatment (Figure 1: diabetes, 3.19 ± 1.14 vs. 4.85 ± 1.23 μ g/ml, $p<0.001$) compared to hypertensive patients without type 2 diabetes that did not have the disease (Figure 1).

Table 2. Blood pressure and angiopathy-related markers in controls and hypertensive patients

	Control	Hypertensive (n=82)	
		DM (-)	DM (+)
n	25	43	39
Sys. BP, mmHg	145 ± 19	187 ± 13 ****	185 ± 16 ****
Dias. BP, mmHg	82 ± 13	95 ± 11 ***	98 ± 10 ****
PDMP, U/ml	5.2 ± 0.6	11.6 ± 2.7 *	17.1 ± 5.6 **
RANTES, ng/ml	55.7 ± 3.8	73.1 ± 12.2 *	88.8 ± 20.4 **
MCP-1, pg/ml	388 ± 87	430 ± 104	492 ± 120 *
sE-selectin, ng/ml	42 ± 10	48 ± 19 *	62 ± 22 ***
adiponectin, μ g/ml	5.93 ± 1.12	4.71 ± 0.85	3.19 ± 1.14 **

Sys. BP: systolic blood pressure; Dias. BP: diastolic blood pressure;
PDMP: platelet-derived microparticle; RANTES: regulated on activation normally T-cell expressed and secreted MCP-1: monocyte chemotactic peptide-1; sE-selectin: soluble E-selectin
Values are shown as mean ± SD.
* $p < 0.05$ ** $p < 0.01$ *** $p < 0.001$ **** $p < 0.0001$

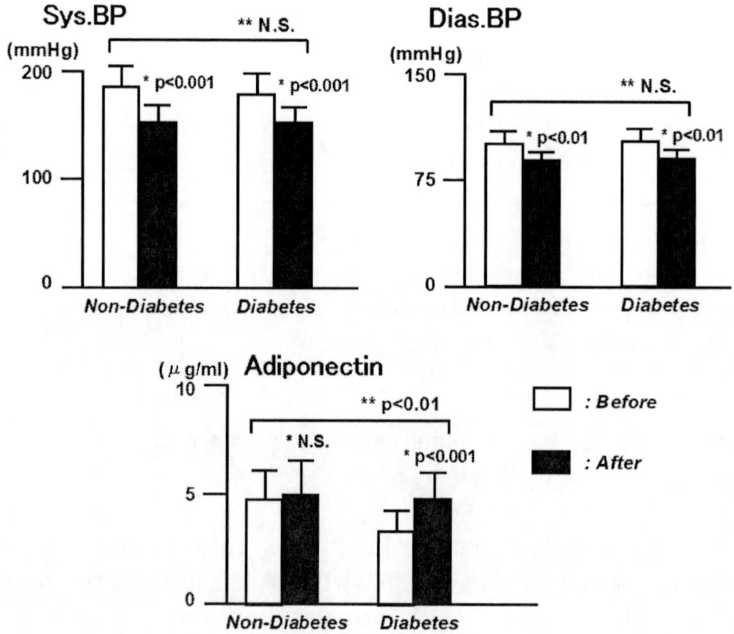

Figure 1. Changes in blood pressure and adiponectin levels before and after administration of LST/HCTZ to hypertensive patients with and without type 2 diabetes. Sys.BP: systolic blood pressure; Dias.BP: diastolic blood pressure. Before: LST/HCTZ treatment (-) After: LST/HCTZ treatement (+). Bars are shown as mean ± S.D. N.S.: not significant. *: Student's t-test (Before vs. After) **: ANOVA analysis (non- diabetes vs. diabetes).

However, total cholesterol, triglyceride, hemoglobin A1c, fasting glucose, and body weight did not exhibit an differences before or after LST/HCTZ administration (data not shown). Levels of angiopathy-related markers were significantly decreased after LST/HCTZ administration in the type 2 diabetes positive group (Figure 2: PDMP, 17.1 ± 5.6 vs. 14.4 ± 3.7 U/ml, p<0.05; RANTES: 88.8 ± 20.4 vs. 77.3 ± 15.2 ng/ml, p < 0.05; MCP-1, 492 ± 120 vs. 464 ± 103 pg/ml, p<0.05; sE-selectin, 62 ± 22 vs. 55 ± 16 ng/ml, p<0.05), but not in the type 2 diabetes negative group (Figure 2).

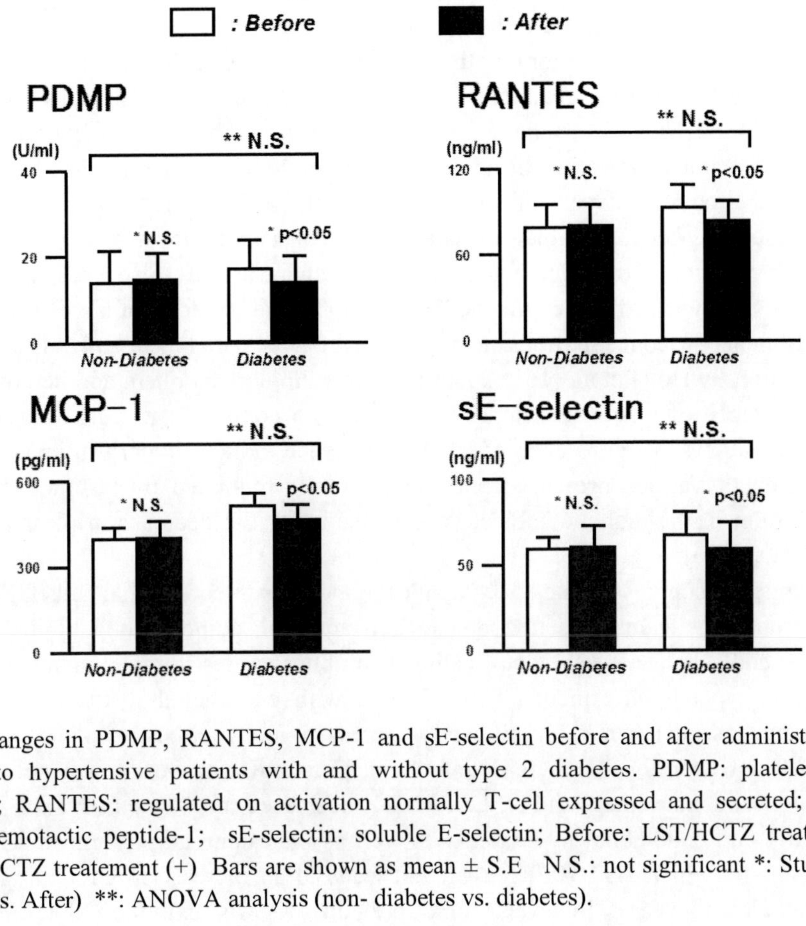

Figure 2. Changes in PDMP, RANTES, MCP-1 and sE-selectin before and after administration of LST/HCTZ to hypertensive patients with and without type 2 diabetes. PDMP: platelet-derived microparticle; RANTES: regulated on activation normally T-cell expressed and secreted; MCP-1: monocyte chemotactic peptide-1; sE-selectin: soluble E-selectin; Before: LST/HCTZ treatment (-) After: LST/HCTZ treatement (+) Bars are shown as mean ± S.E N.S.: not significant *: Student's t-test (Before vs. After) **: ANOVA analysis (non- diabetes vs. diabetes).

DISCUSSION

The present study found that the levels of angiopathy-related markers were higher in hypertensive patients with type 2 diabetes than in those of normotensive controls, and that the administration of LST/HCTZ significantly decreased the concentrations of these markers in patients with type 2 diabetes. In addition, there were no changes in patients that did not have type 2 diabetes. In the present study, hemoglobin A1c, fasting glucose, and body weight did not exhibit any differences before or after LST/HCTZ administration. However, glycemia must be a major contributor to the differences seen between diabetic and non-diabetic individuals. Therefore, the effect of these factors for angiopathy-related markers should be investigated in a different study.

Platelet can synthesize procoagulants which are tissue factors [33-35]. PDMPs are also functionally procoagulant factors because they display surface tissue factors [33]. Alternatively, it is possible to consider PDMPs as markers of platelet activation [7,9,40,42]. Recently, we reported that PDMP concentrations are particularly high in diabetic patients with nephropathy and suggested that the PDMP concentration may be a marker of nephropathy progression in type 2 diabetes [9]. In the present study, PDMPs of diabetic patients with nephropathy exhibited similar findings. The ability of LST/HCTZ to inhibit PDMP generation provides additional evidence for the anti-atherosclerotic and -angiopathic action of this drug. In addition, both systolic and diastolic blood pressure significantly decreased after LST/HCTZ treatment, irrespective of diabetic complications. Therefore, it is possible that LST/HCTZ may have beneficial effects on the vascular system such as the inhibition of platelet activation in addition to its antihypertensive action. Additionally, the reduction in circulating levels of adhesion molecule such as sE-selectin after LST/HCTZ administration provides confirmatory data, and is consistent with a reduction in endothelial dysfunction. The anti-atherogenic properties of adiponectin are also very important in the pathogenesis of type 2 diabetes. In the present study, we observed a significant decrease in plasma adiponectin concentrations in patients with type 2 diabetes, and this finding agrees with the report by Hotta et al. [16]. Abnormalities in lipid metabolism and hemostatic factors, and the presence of insulin resistance are thought to contribute to atherosclerotic vascular damage in diabetes. The presence of insulin resistance induces hyperinsulinemia and affects the secretion of various proteins from adipose tissue regulated by insulin. Thus, elevated plasma insulin in the diabetic subject is responsible for the decrease in plasma adiponectin concentrations [36].

Adiponectin suppressed the attachment of monocytes to endothelial cells [18] and plays a role in protection against vascular damage. In addition, adiponectin inhibitis tissue factor expression and enhances tissue factor pathway inhibitor expression in human endothelial cells [37]. Thus, hypoadiponectinemia is associated with endothelial dysfunction [38]. In the present study, sE-selectin is significantly increased in hypertensive patients with type 2 diabetes. One reason for this increase may be that monocyte activation was not controlled by hypoadiponectinemia. Interestingly, LST/HCTZ decreased monocyte activation by RANTES and MCP-1 and endothelial dysfunction as well as increasing adiponectin levels. The exact mechanisms by which LST/HCTZ treatment leads to an increase in circulating adiponectin levels remains unclear. However, some previous reports exhibited that angiotensin II inhibited the differentiation of preadipocytes, resulting in hypoadiponectinemia [39,40]. Our results support these findings. In addition, hypoadiponectinamia also seems to cause platelet activation. Levels of nitric oxide, which regulate platelet activation, decrease in hypoadiponectinemia since adiponectin stimulates the production of nitric oxide in vascular endothelial cells [19,20], resulting in platelet activation at low concentrations of nitric oxide. Finally, it is thought that platelet activation enhances the adhesion of monocytes and endothelial cells [41]. Regrettably, we could not investigate the relationship between activated platelets and LST/HCTZ using *in vitro* experimental system in the present study. However, our previous reports demonstrated that type 2 diabetes itself could cause platelet hyperaggregability and activated platelets were possible to cause the monocyte activation

[7,8,42]. Further examination is required to elucidate the mechanism of the hypoadiponectinemia improvement by LST/HCTZ.

In conclusion, LST/HCTZ directly or indirectly inhibited the expression of cell adhesion molecules and platelet/monocyte activation for the hypertensive and diabetic patients. These results suggest angiotensin II receptor blockade (LST/HCTZ) may be beneficial as an anti-atherosclerotic -angiopathic therapy in patients with type 2 diabetes in addition to its anti-hypertensive action.

ACKNOWLEDGMENTS

This study was partly supported by a grant from the Japan Foundation of Neuropsychiatry and Hematology Research, a Research Grant for Advanced

Medical Care from the Ministry of Health and Welfare of Japan, and a Grant (13670760 to S.N.) from the Ministry of Education, Science and Culture of Japan.

REFERENCES

[1] Davis TM, Cull CA, Holman RR. Relationship between ethnicity and glycemic control, lipid profiles, and blood pressure during the first 9 years of type 2 diabetes: UK Prospective Diabetes Study (UKPDS 55). *Diabetes Care* 2001; **24:** 1167-74.

[2] Lopes-Virella MF, Virella G. Immune mechanisms of atherosclerosis in diabetes mellitus. *Diabetes* 1992; 41 (Suppl.2): 86-91.

[3] Carmassi F, Morale M, Puccetti R, De Negri F, Monzani F, Navalesi R, et al. Coagulation and fibrinolytic system impairment in insulin dependent diabetes mellitus. *Thromb Res* 1992; **67:** 643-54.

[4] Steiner M, Reinhardt KM, Krammer B, Ernst B, Blann AD. Increased levels of soluble adhesion molecules in type 2 (non-insulin dependent) diabetes mellitus are independent of glycaemic control. *Thromb Haemost* 1994; **72:** 979-84.

[5] Cominacini L, Pasini AF, Garbin U, Campagnola M, Davoli A, Rigoni A, et al. E-selectin plasma concentration is influenced by glycaemic control in NIDDM patients: possible role of oxidative stress. *Diabetologia* 1997; **40:** 584-9.

[6] Uusitupa MI, Niskanen LK, Siitonen O, Voutilainen E, Pyorala K. Ten years cardiovascular mortality in relation to risk factors and abnormalities in lipoprotein composition in type 2 (non-insulin-dependent) diabetic and non-diabetic subjects. *Diabetologia* 1993; **36:** 1175-84.

[7] Nomura S, Ozaki Y, Ikeda Y. Function and role of microparticles in various clinical settings. *Thromb Res* 2008;**123**: 8-23.

[8] Nomura S, Inami N, Shouzu A, Urase F, Maeda Y. Correlation and association between plasma platelet-, monocyte- and endothelial cell-derived microparticles in hypertensive patients with type 2 diabetes mellitus. *Platelets* 2009;**20**: 406-414.

[9] Nomura S, Shouzu A, Taomoto K, Togane Y, Goto S, Ozaki Y, et al. Assessment of an ELISA kit for platelet-derived microparticles by joint research at many institutes in Japan. *J Atheroscler Thromb* 2009:**16**; 878-887.

[10] Rollins BJ, Yoshimura T, Leonard EJ, Pober JS. Cytokine-activated human endothelial cells synthesize and secrete a mono- chemoattractant, MCP-1/JE. *Am J Pathol* 1990; **136:** 1229-33.

[11] Colotta F, Borre A, Wang JM, Tattanelli M, Maddalena F, Polentarutti N, et al. Expression of a monocyte chemotactic cytokine by human mononuclear phagocytes. *J Immunol* 1992; **148:** 760-5.

[12] Ihm CG, Park JK, Hong SP, Lee TW, Cho BS, Kim MJ, et al. A high glucose concentration stimulates the expression of monocyte chemotactic peptide 1 in human mesangial cells. *Nephron* 1998; **79:** 33-7.

[13] Fries JWU, Williams AJ, Atkins RC, Newman W, Lipscomb MF, Collins T. Expression of VCAM-1 and E-selectin in an in vivo model of endothelial activation. *Am J Pathol* 1993; **143**: 725-37.

[14] Ouchi N, Kihara S, Arita Y, Okamoto Y, Maeda K, Kuriyama H, et al. Adiponectin, an adipocyte-derived plasma protein, inhibits endothelial NF-kappa B signaling through a cAMP-dependent pathway. *Circulation* 2000; **102**: 1296-301.

[15] Arita Y, Kihara S, Ouchi N, Takahashi M, Maeda K, Miyagawa J, et al. Paradoxical decrease of an adipose-specific protein, adiponectin, in obesity. *Biochem Biophys Res Commun* 1999; **257**: 79-83.

[16] Hotta K, Funahashi T, Arita Y, Takahashi M, Matsuda M, Okamoto Y, et al. Plasma concentrations of a novel, adipose-specific protein, adiponectin, in type 2 diabetes patients. *Artherioscler Thromb Vasc Biol* 2000; **20**: 1595-9.

[17] Weyer C, Funahashi T, Tanaka S, Hotta K, Matsuzawa Y, Pratley RE, et al. Hypoadiponectinemia in obesity and type 2 diabetes: close association with insulin resistance and hyperinsulinemia. *J Clin Endocrinol Metab* 2001; **86**: 1930-5.

[18] Ouchi N, Kihara S, Arita Y, Maeda K, Kuriyama H, Okamoto Y, et al. Novel modulator for endothelial adhesion molecules: adipocyte-derived plasma protein, adiponectin. *Circulation* 1999; **100**: 2473-6.

[19] Chen H, Montagnani M, Funahashi T, Shimomura I, Quon MJ. Adiponectin stimulates production of nitric oxide vascular endothelial cells. *J Biol Chem* 2003; **278:** 45021-6.

[20] Hattori Y, Suzuki M, Hattori S, Kasai K. Globular adiponectin upregulates nitric oxide production in vascular endothelial cells. *Diabetologia* 2003; **46:** 1543-9.

[21] Ingelfinger JR, Dzau VJ. Molecular biology of renal injury: emphasis of the role of the rennin-angiotensin system. *J Am Soc Nephrol* 1991; **2:** S9-S20.

[22] Lindpainter K, Ganten D. The cardiac Renin-angiotensin system: an appraisal of present experimental and clinical evidence. *Circ Res* 1991; **68:** 905-21.

[23] Bech Laursen J, Rajagopalan S, Galis Z, Tarpey M, Freeman BA, Harrison DG. Role of superoxide in angiotensin II-induced but not catecholamine-induced hypertension. *Circulation* 1997; **95:** 588-93.

[24] Grafe M, Auch-Schwelk W, Zakrzewicz A, Regitz-Zagrosek V, Bartsch P, Graf K, et al. Angiotensin II-induced leukocyte adhesion on human coronary endothelial cells in mediated by E-selectin. *Circ Res* 1997; **81:** 804-11.

[25] Parving HH, Lehnert H, Brochner-Mortensen J, Gomis R, Andersen S, Arner P, et al. The effect of irbesartan on the development of diabetic nephropathy in patients with type 2 diabetes. *N Engl J Med* 2001;**345**: 870-878

[26] Viberti G, Wheeldon NM. Microalbuminuria reduction with valsartan in patients with type 2 diabetes mellitus: a blood pressure-independent effect. *Circulation* 2002;**106**: 672-678.

[27] Garg JP, Bakris GL. Microalbuminuria: marker of vascular dysfunction, risk factor for cardiovascular disease. *Vasc Med* 2002;**7**: 35-43.

[28] Gerstein HC, Mann JF, Yi Q, Zinman B, Dinneen SF, Hoogwerf B, et al. Albuminuria and risk of cardiovascular events, death, and heart failure in diabetic and nondiabetic individuals. *J Am Med Assoc* 2001:**286**; 421-426

[29] Mancia G, De Backer G, Dominiczak A, Cifkova R, Fagard R, Germano G, et al. 2007 ESH-ESC practice guidelines for the management of arterial hypertension: ESH-ESC Task Force on the management of arterial hypertension. *J Hypertens* 2007;**25**: 1751-1762.

[30] Mogensen CE, Viberti G, Halimi S, Ritz E, Ruilope L Jermendy G, et al. Effect of low-dose peridopril/indapamide on albuminuria in diabetes: preterax in albuminuria regression: PREMIER. *Hypertension* 2003;**41**: 1063-1071.

[31] The National High Blood Pressure Education Program Working Group National High Blood Pressure Education Program Working Group Report on Hypertension in diabetes. *Hypertension* 1994; **23:** 145-58.

[32] The Expert Committee on the Diagnosis and Classification of Diabetes Mellitus. Report of the expert commitee on the diagnosis and classification of diabetes mellitus. *Diabetes Care* 1997; **20:** 1183-97.

[33] Muller I, Klocke A, Alex M, Kotzsch M, Luther T, Morgenstern E, et al. Intravascular tissue factor initiates coagulation via circulating microvesicles and platelets. *FASEB J* 2003;**17**: 476-478.

[34] Perez-Pujol S, Aras O, Lozano M, Cocking-Johnson D, Key NSWhite JG, et al. Stored platelets contain residual amounts of tissue factor: evidence from studies on platelet concentrates stored for prolonged periods. *Transfusion* 2005;**45**: 572-579.

[35] Panes O, Matus V, Saez CG, Quiroga T, Pereria J, Mezzano D. Human platelets synthesize and express functional tissue factor. *Blood* 2007; **109**: 5242-5250.

[36] Mohlig M, Wegewitz U, Osterhoff M, Isken F, Ristow M, Pfeiffer AF, et al. Insulin decreases human adiponectin plasma levels. *Horm Metab Res* 2002; **34:** 655-8.

[37] Chen YJ, Zhang LQ, Wang GP, Zeng H, Lu B, Shen XL, Jiang ZP. Adiponectin inhibits tissue factor expression and enhances tissue factor pathway inhibitor expression in human endothelial cells. *Thromb. Haemost* 2008; 100: 291-300.

[38] Shimabukuro M, Higa N, Asahi T, Oshiro Y, Takasu N, Tagawa T, et al. Hypoadiponectinemia is closely linked to endothelial dysfunction in man. *J Clin Endocrinol Metab* 2003; **88:** 3236-40.

[39] Sharma AM, Janke J, Gorzelniak K, Engeli S, Luft FC. Angiotensin blockade prevents type 2 diabetes by formation of fat cells. *Hypertension* 2002; **40:** 609-11.

[40] Furuhashi M, Ura N, Higashiura K, Murakami H, Tanaka M, Moniwa N, et al. Blockade of the rennin-angiotensin system increases adiponectin concentrations in patients with essential hypertension. *Hypertension* 2003; **42:** 76-81.
[41] Nomura S, Tandon NN, Nakamura T, Cone J, Fukuhara S, Kambayashi J. High-shear-stress-induced activation of platelets and microparticles enhances expression of cell adhesion molecules in THP-1 and endothelial cells. *Atherosclerosis* 2001;**158:** 277-87.
[42] Nomura S, Shouzu A, Omoto S, Nishikawa M, Iwasaka T. Long-term treatment with nifedipine modulates procoagulant marker and C-C chemokine in hypertensive patients with type 2 diabetes mellitus. *Thromb Res* 2005; **115**: 277-85.

In: Adiponectin: Production, Regulation and Roles in Disease ISBN: 978-1-61942-165-3
Editor: Masayoshi Yamaguchi

Chapter IX

First Trimester Serum Adiponectin Levels in Women Later Developing Mild Pre-Eclampsia and Their Comparison with Other Hormones

Nick A. Bersinger[1]
University of Berne, Switzerland

Abstract

Adiponectin is an adipose tissue derived protein which, in spite of its high circulating concentrations, has not been extensively studied. Its serum concentrations were reported negatively to correlate to body mass and, in parallel, to the circulating leptin levels. It was also found to inversely correlate with insulin resistance and, in pregnancy when compared to healthy pregnant women, to be decreased in the presence of gestational diabetes. In pre-eclampsia (PE) after its onset, circulating adiponectin was reported to increase in the majority but not in all studies. When measured in the first trimester, adiponectin levels were found to be decreased in women later developing pre-eclampsia in one study, while another reported increased concentrations in early-onset and unchanged levels in late-onset pre-eclampsia.

A retrospective analysis was performed with 50 patients who subsequently developed mild, late-onset pre-eclampsia which is the most frequent manifestation of the disease. Sera were obtained between 11+2 and 13+4 weeks of gestation, and each of them was matched with two control sera for gestational age, maternal age, maternal pre-pregnancy weight, and duration of storage. Total circulating adiponectin was determined by microplate ELISA. The serum levels of a panel of hormones and growth factors in the same samples were available either from routine screening or from previous studies performed in our laboratory. Adiponectin concentrations were cross-sectionally compared between the cases and the matched healthy controls by Mann-Whitney U test, and correlations between adiponectin and other hormones and growth factors were studied using Spearman's Rank Correlation analysis.

[1] E-mail address: nick.bersinger@dkf.unibe.ch

Concentrations of total adiponectin did not differ in our sample of first trimester sera between women developing mild, late-onset pre-eclampsia and matched controls. The medians were 10.8 (range 2.1 – 21.4) in the cases and 9.8 (range 2.9 – 22.9) μg/mL, respectively. Adiponectin levels were negatively correlated with the maternal pre-pregnancy weight and with the serum concentrations of leptin and C-reactive protein. Positive correlations were observed with the free β-subunit of hCG, PAPP-A, soluble endoglin, the soluble form of Flt-1, and inhibin A but not with activin A.

Adiponectin does not seem to be a marker for early detection of pre-eclampsia, or at least not for its mild, more frequent form. However, it is interesting to note that it is predominantly the proteins originating from the placenta that show positive correlations with adiponectin.

Keywords: pre-eclampsia, pregnancy

INTRODUCTION

Pre-eclampsia (PE) is a major pregnancy disorder defined by hypertension and proteinuria and which occurs with an incidence of around 5%. Its pathogenesis is still poorly understood since it is a complex disease. Although it can only be treated efficiently by the delivery of the placenta, the pathology and subsequent abnormal development originates in early pregnancy through restricted trophoblast invasion and insufficient adaptation of the spiral arteries, which in turn is leading to placental hypoxia. The maternal response is characterised not only by increased blood pressure and proteinuria but also by oedema, abnormal blood clotting and abnormal liver and renal function. These symptoms, however, are not seen until later in pregnancy and are associated with endothelial dysfunction in an overreacting inflammatory response [1,2]. After the onset of these maternal symptoms the serum concentrations of soluble endothelial markers are increased [3], together with the placentally derived inhibin A, activin A [4] and pregnancy-associated plasma protein A (PAPP-A) [5,6]. The mechanisms leading to these elevated amounts released may be the same for the maternal and the placental compartment since we found a positive correlation between PAPP-A and the soluble endothelial (sE-) selectin [6], but a non-placental production of PAPP-A cannot be ruled out [7], particularly in normal, healthy pregnancy [8].

In the quest of early diagnosis or at least for screening purposes to identify pregnancies at risk it became increasingly important to identify circulating markers whose serum concentrations in the first or second trimester of pregnancy would differ between women with normally proceeding pregnancies and those who subsequently would develop pre-eclampsia. Placental factors play an early role, and a logical approach consisted in searching for serum markers from this source for distinguishing between high- and low-risk pregnancies regarding PE. Inhibin A and activin A have been the most intensely studied factors in this group of molecules, and raised early pregnancy inhibin A levels have been reported in women who later developed severe PE [9,10]. PAPP-A, unexpectedly, was found to be reduced in early pregnancy in PE patients [11], an observation we have confirmed in a cross-sectional study [12]. A longitudinal investigation revealed that this reduction compared to healthy pregnancy levels was disappearing with advancing gestation and then leading up to the previously demonstrated increased levels after onset of the maternal symptoms [6]. To our knowledge the only other marker found to be reduced in the serum of pre-eclamptic patients in the third

[12,13], the second [12,14], but apparently not in the first [15,16] trimester is placenta growth factor (PLGF). The concentrations of pregnancy-specific β1-glycoprotein (SP1) and placental lactogen (hPL) do not seem to be related to the risk of PE, but information on these markers is scarce and comparatively old [17,18].

Studies on the role and usefulness for screening purposes of pro-inflammatory cytokines, such as the interleukins (IL)-6, IL-8, and tumour necrosis factor-α (TNF-α) in early pregnancy are scarce, for reasons of their pleiotropic behaviour and also due to limitations as a consequence of the often insufficient sensitivities of many of the respective immunoassays. Moreover, the small available body of information with regards to pre-eclampsia is inconsistent. C-reactive protein (CRP), an inflammatory marker, does not pose sensitivity problems with a new assay method [19], but the specificity may be insufficient; it has been studied in the context of pre-eclampsia and increased serum levels were found in the third trimester in patients. Statistical significance of this finding, however, was lost after correcting for the body mass index [20].

Leptin is the product of the *ob* gene [21], synthesised and released into the circulation by peripheral adipocytes; its serum levels strongly correlate with the body mass index. In pregnancy, leptin is also produced by the placenta [22], resulting in increased serum concentrations as a function of advancing gestation until the mid-third trimester [23]. In patients suffering from pre-eclampsia serum leptin levels were found to be increased after the onset of the maternal symptoms [24]. In earlier pregnancy, i.e. before the onset of symptoms, leptin was shown not to be elevated in the first [25] or even in the second [26] trimester of subsequently pre-eclamptic gestations. Adiponectin (Acrp30, 244 amino acids) is another adipose tissue derived protein [27] which, in spite of its high circulating concentrations, has not been extensively studied. Paradoxically and in contrast to the circulating leptin levels, its serum concentrations were reported to be negatively correlated to body mass [28]. Similarly adiponectin was found to be inversely correlated with insulin resistance [29] and, in pregnancy when compared to healthy pregnant women, to be decreased in the presence of gestational diabetes [30]. In pre-eclampsia after the onset of the maternal symptoms, circulating adiponectin was reported to be increased in the majority but not in all studies [31,32]. When measured in the first trimester, adiponectin levels were found to be decreased in women later developing pre-eclampsia over normotensive controls in one study [33], while another reported increased concentrations in early-onset and unchanged levels in late-onset pre-eclampsia [34].

The aim of the project described in this chapter was to investigate, in a carefully matched case-control study, the association between the serum total adiponectin concentration in the first trimester of pregnancy and the later occurrence of (essentially mild, always late-onset) pre-eclampsia. Moreover, the correlations between the adiponectin levels and the concentrations of a panel of other hormones and growth factors of placental and non-placental origin were analysed in the overall population as well as in our group of PE cases.

Patients, Materials and Methods

Patients and Matched Controls

A retrospective analysis on banked sera from a previous study [16] and with complete clinical information was performed. Fifty patients who subsequently developed mild, late-onset (after 34+5 weeks of gestation) pre-eclampsia were included. Cases of severe, early-onset pre-eclampsia, HELLP syndrome pregnancies and other hypertensive or renal non-PE disorders were excluded. Patient sera were obtained in the narrow late first trimester interval between 11+2 and 13+4 weeks of gestation. PE was defined as a diastolic blood pressure above 90 mm Hg or a systolic pressure above 140 mm Hg together with proteinuria of at least 300 mg in a 24-hour period [35]. For these women the referring obstetricians were reporting back to the private analytical laboratory where the sera had been collected (MCL, Düdingen, Switzerland), providing the gestational age at birth and the birthweight of the baby. The case group (mild, late-onset PE) was characterised by a mean gestational age at birth of 38+2 weeks, a child birth weight of 2975 g, and a placental weight of 551 g. Fetal sex outcome was 20 boys and 26 girls (plus 4 without available information). Each serum from a PE patient was matched with two control sera, from pregnant women who did not develop PE, for gestational age (by the day, from ultrasound measurements), maternal age, duration of storage, and the pre-pregnancy maternal weight. This latter parameter of matching procedure was decided to be necessary because this was an adipokine study and not only leptin, but also CRP levels in the serum were reported to be positively correlated to the body mass [36]. The sera were stored at -80°C until adiponectin was determined in batches.

Measurement of Adiponectin

Total adiponectin (all isoforms combined) was determined using a microplate enzyme immunometric assay developed in our laboratory. Briefly, Nunc Maxisorp® plates were coated with the monoclonal capture antibody (MAB10651, R&D Systems, Oxford, England) solution (100 µL/well at 1.0 µg/mL in PBS, pH 7.3) overnight at 4°C. Then the solution was aspirated and the wells blocked by the immediate addition (no washing step) of bovine serum albumin (BSA, Sigma, St. Louis, USA), 250 µL/well at 0.5 mg/mL in PBS for two hours at 30°C. After washing using PBS containing Tween-20 (Sigma) at 0.1% v/v and an automatic plate washer (three cycles), standards and patient sera were added to the wells (100µL) in duplicate. Sera were diluted 1:5000 in PBS containing BSA at 0.1% w/v, and the recombinant adiponectin standard (obtained from Biovendor, Brno, Czech Republic) was used in serial 1:2 dilution steps between 8.0 and 0.125 ng/mL. Incubation was for two hours at 28°C in a dry plate incubator without shaking. After washing (as above), the biotinylated detection antibody was added at a concentration of 2 µg/mL in the same diluent (BSA in PBS) and the plates incubated for another two hours at 28°C. After thorough washing (4 cycles of PBS-Tween-20), the streptavidin-horseradish peroxidase (HRP) conjugate (R&D Systems, Part 890803) was added at a 1:200 dilution as recommended by the manufacturer and the plates incubated for 20 minutes. Then they were washed and the substrate added (TMB single solution ready to use, Invitrogen, Frederick, USA). Incubation was at room temperature and in the dark, for

10 – 25 minutes depending on the development of the (blue) colour in the wells, then the reaction was stopped by the addition of in the same timed sequence. Optical density was read at 450 nm (yellow) against a background of 620 nm in an ASYS Expert Plus microplate reader (Biolabo, Châtel St-Denis, Switzerland). The functional sensitivity of the assay was 0.05 ng/mL.

Determination of Other Hormones and Markers

The serum levels of a panel of hormones and growth factors in the same samples were available either from routine first-trimester fetal trisomy screening (free β-HCG, PAPP-A), performed in the MCL laboratory, or from previous studies performed in our laboratory and involving other placental proteins such as pregnancy-specific β1 glycoprotein (SP1) [37], human placental lactogen (hPL) [38] or placenta growth factor (PLGF) [12], or non-placental markers such as C-reactive protein (CRP) [19], interleukin-8 (IL-8) [39], soluble endoglin and soluble Flt-1 [40]. Dimeric inhibin A (ultra-sensitive) and activin A were quantified previously in the study sera using amplified alkaline phosphatase based kits from Serotec, Oxford, England after a denaturation step. Leptin was determined with a double-antibody enzyme immunometric assay developed in our laboratory and described elsewhere [41].

The serum levels of PAPP-A and free β-hCG in the same samples were available from the routine first-trimester Down syndrome screening programme. To this purpose a fully automated method (Kryptor©, CIS Bio, France) was used.

Descriptive Statistical Analysis

Means, standard deviations, medians and ranges were tabulated, and the adiponectin concentrations were cross-sectionally compared between the cases of late-onset pre-eclampsia and the matched healthy controls by Mann-Whitney U test. Correlations between adiponectin and the above mentioned hormones and growth factors were studied using Spearman's Rank Correlation analysis.

Results

The first trimester serum concentrations of total adiponectin did not differ in our study population between women developing mild, late-onset pre-eclampsia and matched controls. The medians were 10.8 (range 2.1 – 21.4) in the cases and 9.8 (range 2.9 – 22.9) μg/mL, respectively. Amongst the other markers tested in this project and in previous analyses only inhibin A, activin A and soluble endoglin were found to be significantly increased in the cases over the controls; for soluble fms-like tyrosine kinase 1 higher levels were also observed but statistical significance was not quite reached. Details are given in Table 2. The soluble form of endothelial selectin did not differ between the two groups (not shown).

Table 1. Demographic and clinical data of study groups. Values are means ± S.D.

	Cases (N=50)	Controls (N=100)
Age (years)	31.0 ± 4.8	31.4 ± 4.4
Maternal weight, kg	72.6 ± 16.4	72.4 ± 14.7
Gestational age at sampling, days	86.1 ± 4.9	86.2 ± 4.7
Nuchal translucency thickness, mm	1.47 ± 0.51	1.38 ± 0.52

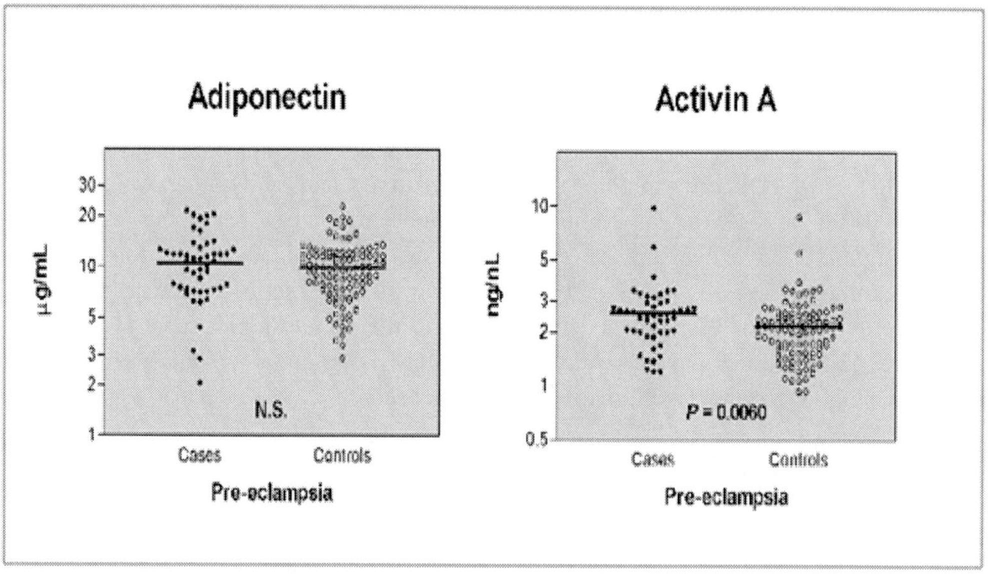

Figure 1. First trimester serum adiponectin (left panel) and Inhibin A (right panel) concentrations in patients later developing mild, late onset per-eclampsia (N=50) and control, healthy pregnancies (N=100) matched for gestational age, pre-pregnancy maternal weight, and storage time. The difference for activin A is statistically significant, but with a large variation. Please not the logarithmic scales.

When all 150 pregnancies were taken together, adiponectin levels were significantly ($P<0.05$) and negatively correlated not only with the maternal pre-pregnancy weight and the serum leptin concentration (Fig.2, left and centre panels), but also with C-reactive protein. On the other hand, significant positive correlations were observed with the free β-subunit of hCG, PAPP-A (Fig.2, right panel), and with soluble endoglin. Inhibin A and the soluble form of Flt-1 were also positively associated with adiponectin but statistical significance was not quite reached ($P=0.06$). For activin A, however, no trend could be made out. The results for these correlation analyses are shown in Table 3. When the case group (N=50) was analysed separately in this respect the negative correlation with the leptin levels became stronger and so did the positive association between adiponectin with s-Flt1, but all other correlations mentioned above for the whole population disappeared. No association between the adiponectin levels and the maternal age was observed.

Table 2. First trimester adiponectin and other marker concentrations in cases of later developing mild pre-eclampsia (N=50) and matched, healthy controls (N=100). For abbreviations see Methods. Values are medians and range. *P* values were obtained using the Mann-Whitney U test. Significant differences between groups are marked in bold print.

	Cases (N=50)	Controls (N=100)	*P*
Adipokines			
Adiponectin (μg/mL)	10.8 (2.1 – 21.4)	9.8 (2.9 – 22.9)	0.3367
Leptin (ng/mL)	37.4 (6.4 – 119.3)	30.6 (7.3 – 122.5)	0.2388
Other markers			
Free β-hCG (mIU/mL)	35.4 (11.7 – 104.1)	36.3 (10.4 – 137.9)	0.7630
PAPP-A (mIU/mL)	2.32 (0.49 – 14.29)	3.08 (0.54 – 9.46)	0.3925
SP1 (mg/mL)	8.90 (3.26 – 40.90)	9.70 (1.93 – 28.06)	0.9650
hPL (mg/mL)	0.633 (0.122 – 1.231)	0.679 (0.162 – 1.416)	0.3624
PLGF (ng/mL)	0.150 (0.068 – 2.409)	0.142 (0.056 – 0.275)	0.7649
Inhibin A (ng/mL)	0.189 (0.078 – 0.748)	0.149 (0.026 – 0.585)	**0.0320**
Activin A (ng/mL)	2.47 (1.20 – 9.76)	2.11 (0.93 – 8.72)	**0.0060**
s-Eng (ng/mL)	5.51 (3.77 – 8.70)	4.80 (3.08 – 7.53)	**0.0094**
s-Flt1 (pg/mL)	1503 (625 – 3843)	1354 (525 – 6212)	0.0912
CRP (mg/mL)	6.68 (0.55 – 31.38)	6.32 (0.22 – 127.14)	0.9603
IL-8 (pg/mL)	7.12 (1.78 – 43.20)	6.72 (1.73 – 28.00)	0.2881

DISCUSSION

The serum concentrations of total adiponectin, determined in a comparatively narrow and well controlled time bracket in early pregnancy, did not differ in our study population between the 50 women developing mild, late-onset pre-eclampsia and 100 matched controls. The same was the case for leptin, the other adipose tissue marker, confirming earlier first trimester in comparison to post-onset studies [24,25]; the two were inversely correlated which is in agreement to the literature [28]. Our findings on adiponectin disagree with the results reported earlier on 34 pre-eclampsia patients [33] but confirm the more recently published findings observed in a population of similar size as ours presented in this chapter: serum adiponectin was increased in the group of 30 cases developing early pre-eclampsia but not in the 60 women who later presented the mild, late onset version of the disease [34]. Moreover, and again in agreement with our findings, no association between the serum levels of adiponectin and PAPP-A was found by Nanda et al. [34]. This also validates the use of our assay method developed in our laboratory. Our observation similarly confirms the finding of O'Sullivan et al. [32] since our sample did not contain women suffering from diabetes.

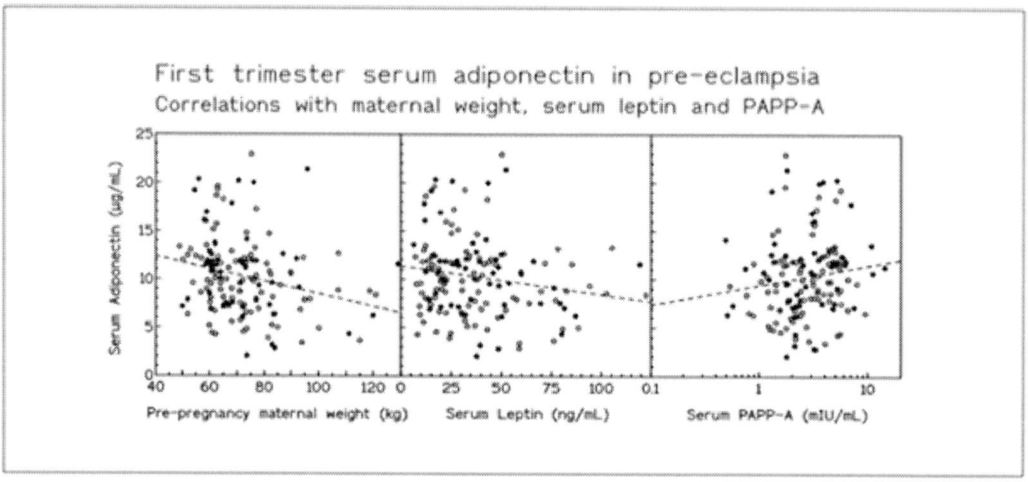

Figure 2. Correlations in first trimester serum between the adiponectin level and the maternal pre-pregnancy weight (left panel) and the concentrations of leptin (centre panel) and pregnancy-associated plasma protein A (right panel). Closed circles, cases later developing mild pre-eclampsia (N=50); open circles, matched control healthy pregnancies (N=100). The dotted lines represent the linear regression curves over all 150 women; all correlations shown, where negative (maternal weight and leptin) or positive (PAPP-A), are statistically significant (P<0.05).

Only inhibin A, activin A and soluble endoglin were found to be increased in the cases over the matched controls, and for soluble fms-like tyrosine kinase 1 higher levels were also observed but statistical significance was just not quite reached (P=0.0912, Table 2). We have also examined the performance of the product adiponectin x activin A since the latter marker was both the most strongly increased one in our PE group over the controls and at the same time unrelated to adiponectin. Not surprisingly, a statistically significant difference between our PE and control groups was found (P=0.0172), but this was less pronounced that it was for activin A alone (P=0.0060).

Leptin, CRP and IL-8 are related to inflammatory processes [42] such as pre-eclampsia, but this may play a role only later in gestation. Our correlations with adiponectin presented here are negative or absent, which further speaks against an usefulness of this molecule in screening to identify pregnancies carrying a risk to later develop PE. On the other hand, the exclusively placental markers SP1 and free β-hCG are also no indicators for an increased PE risk while inhibin A and activin A, which are partially but not exclusively produced by the placenta, are capable of distinguishing between the healthy and the at-risk groups.

Adiponectin thus does not seem to be a good molecule for the early detection of pre-eclampsia or at least not for its mild, more frequent form. It may be interesting to investigate whether the high molecular weight isoform of the substance would be a better predictor than total adiponectin for pre-eclampsia as it has been shown for other pathologies [43,44]. Nevertheless, it seems unlikely that a single ‘miracle’ serum marker in early pregnancy will be available in the near future for reliably predicting which women would subsequently develop and which of them won't carry that risk.

Table 3. Correlations by (non-parametric) Spearman R ank Analysis between serum adiponectin and clinical parameters or other serum markers in the first trimester of pregnancy. Values are combined for all women (cases and controls, N=150). For abbreviations see Methods. For markers not listed in this table, no correlations or significant trends towards an association were observed (*P*>0.1). *P* values below 0.05 are marked in bold.

	Positive r	Negative r	Spearman *P*
Maternal age		-0.1094	0.1825
Maternal pre-pregnancy weight		-0.2852	**0.0004**
Leptin		-0.1698	**0.0378**
CRP		-0.1628	**0.0465**
Free β-hCG	0.1635		**0.0463**
PAPP-A	0.1810		**0.0267**
hPL	0.1488		0.0692
s-Eng	0.1778		**0.0370**
Inhibin A	0.1525		0.0625
s-Flt1	0.1617		0.0641
sE-Selectin	0.1961		0.0687

REFERENCES

[1] Redman,C.W., Sacks,G.P. & Sargent,I.L. Pre-eclampsia: an excessive maternal inflammatory response to pregnancy. Am. J. Obstet. Gynecol. 1999; 180: 499-506.

[2] Chaiworapongsa,T., Romero,R., Yoshimatsu,J., Espinoza,J., Kim,Y.M., Park,K. et al. Soluble adhesion molecule profile in normal pregnancy and pre-eclampsia. J. Matern. Fetal Neonat. Med. 2002; 12: 19-27.

[3] Austgulen,R., Lien,E., Vince,G. & Redman,C.W. Increased maternal plasma levels of soluble adhesion molecules (ICAM-1, VCAM-1, E-selectin) in preeclampsia. Eur. J. Obstet. Gynecol. Reprod. Biol. 1997; 71: 53-58.

[4] Muttukrishna,S., Knight,P.G., Groome,N.P., Redman,C.W. & Ledger,W.L. Activin A and inhibin A as possible endocrine markers for pre-eclampsia. Lancet 1997; 349: 1285-1288.

[5] Toop,K. & Klopper,A. Concentration of pregnancy-associated plasma protein A (PAPP-A) in patients with pre-eclamptic toxaemia. Placenta 1981; 4(Suppl.): 167-173.

[6] Bersinger,N.A., Smarason,A., Muttukrishna,S., Groome,N.P. & Redman,C.W. Women with pre-eclampsia have increased serum levels of pregnancy-associated plasma protein A (PAPP-A), inhibin A, activin A, and soluble E-selectin. Hypertens. Preg. 2003; 22: 45-55.

[7] Bersinger,N.A., Altermatt,H.J., Birkhäuser,M.H., Brizot,M.L., Kappeler,A., Muller,M.D. et al. Non-placental production of pregnancy-associated plasma protein A (PAPP-A): old and new evidence. Early Preg. Biol. Med. 1997: 3: 96-101.

[8] Bersinger,N.A., Marguerat,P., Pescia,G. & Schneider,H. Pregnancy-associated plasma protein A (PAPP-A): measurement by highly sensitive and specific enzyme immunoassay, importance of first-trimester serum determinations, and stability studies. Reprod. Fert. Dev. 1995; 7: 1419-1423.

[9] Cuckle,H., Sehmi,I. & Jones,R. Maternal serum inhibin A can predict pre-eclampsia. Brit. J. Obstet. Gynaecol. 1998; 105: 1101-1103.

[10] Sebire,N.J., Roberts,L., Noble,P., Wallace,E. & Nicolaides,K.H. Raised maternal serum inhibin A concentration at 10 to 14 weeks of gestation is associated with pre-eclampsia. Brit. J. Obstet. Gynaecol. 2000; 107: 795-797.

[11] Smith,G.C., Stenhouse,E.J., Crossley,J.A., Aitken,D.A., Cameron,A.D. & Connor,J.M. Early pregnancy levels of pregnancy-associated plasma protein A and the risk of intrauterine growth restriction, premature birth, preeclampsia and stillbirth. J. Clin. Endocrinol. Metab. 2002; 87: 1762- 1767.

[12] Bersinger,N.A. & Odegard,R.A. Second- and third-trimester serum levels of placental proteins in preeclampsia and small-for-gestational age pregnancies. Acta Obstet. Gynecol. Scand. 2004; 83: 37-45.

[13] Torry,D.S., Wang,H.S., Wang,T.H., Caudle,M.R. & Torry,R.J. Preeclampsia is associated with reduced serum levels of placenta growth factor. Am. J. Obstet. Gynecol. 1998; 179: 1539-1544.

[14] Su,Y.N., Lee,C.N., Cheng,W.F., Shau,W.Y., Chow,S.N. & Hsieh,F.J. Decreased maternal serum placenta growth factor in early second trimester and preeclampsia.Obstet. Gynecol. 2001; 97: 898-904.

[15] Ong,C.Y., Liao,A.W., Cacho,A.M., Spencer,K. & Nicolaides,K.H. First-trimester maternal serum levels of placenta growth factor as predictor of preeclampsia and fetal growth restriction. Obstet. Gynecol. 2001; 98: 608-611.

[16] Zwahlen,M., Gerber,S. & Bersinger,N.A. First trimester markers for pre-eclampsia: placental vs. non-placental protein serum levels. Gynecol. Obstet. Invest. 2007, 63: 15-21. *Erratum*: Please note that the units in Fig.1 (page 18) in this publication should read "ng/mL" (instead of "mg/mL") for Inhibin A, Activin A, and PLGF. The units of the other graphs in this Figure are correct.

[17] Chapman,M.G., O'Shea,R.T., Jones,W.R. & Hillier,R. Pregnancy-specific β1-glycoprotein as a screening test for at-risk pregnancies. Am. J. Obstet. Gynecol. 1981; 141: 499-502.

[18] Obiekwe,B.C., Sturdee,D., Cockrill,B.L. & Chard,T. Human placental lactogen in pre-ecplampsia. Brit. J. Obstet. Gynaecol. 1984; 91: 1077-1080.

[19] Wunder,D.M., Kretschmer,R. & Bersinger,N.A. Concentrations of leptin and C-reactive protein in serum and follicular fluid during assisted reproductive cycles. Hum. Reprod. 2005; 20: 1266-1271.

[20] Belo,L., Santos,A., Caslake,A., Cooney,J., Pereira,L., Quintanilha,A. Et al. Neutrophil activation and C-reactive protein concentration in preeclampsia. Hypertens. Preg. 2003; 22: 129-141.

[21] Zhang,Y., Proenca,R., Maffei,M., Barone,M., Leopold,L. & Friedman,J.M. Positional cloning of the mouse obese gene and its human homologue. Nature 1994; 372: 425–432.
[22] Masuzaki,H., Ogawa,Y., Sagawa,N., Hosoda,K., Matsumoto,T., Mise,H. et al. Nonadipose tissue production of leptin: leptin as a novel placenta-derived hormone in humans. Nature Med. 1997; 3: 1029–1033.
[23] Highman,T.J., Friedman,J.E., Huston,L.P., Wong,W.W. & Catalano,P.M. Longitudinal changes in maternal serum leptin concentrations, body composition, and resting metabolic rate in pregnancy. Am. J. Obstet. Gynecol. 1998; 178: 1010–1015.
[24] Bartha,J.L., Romero,R., Escobar,M. & Comino,R. The relationships between leptin and inflammatory cytokines in women with preeclampsia. Br. J. Obstet. Gynaecol. 2001; 108: 1272–1276.
[25] Salomon,L.J., Benattar,C., Audibert,F., Fernandez,H., Duyme,M., Taieb,J. et al. Severe preeclampsia is associated with high inhibin A levels and normal leptin levels at 7 to 13 weeks into pregnancy. Am. J. Obstet. Gynecol. 2003; 189: 1517–1522.
[26] Chan,T.F., Su,J.H., Chung,Y.F., Hsu,Y.H., Yeh,Y.T., Jong,S.B. et al. Amniotic and maternal serum leptin levels in pregnant women who subsequently develop preeclampsia. Eur. J. Obstet. Gynecol. Reprod. Biol. 2003; 108: 50–53.
[27] Iyengar,P. & Scherer,P.E. Adiponectin/Acrp30, an adipocytespecific secretory factor: physiological relevance during development. Pediatr. Diabetes. 2003; 4:32–37.
[28] Weyer,C., Funahashi,T., Tanaka,S., Hotta,K., Matsuzawa,Y., Pratley,R.E., et al. Hypoadiponectin-emia in obesity and type 2 diabetes: close association with insulin resistance and hyperinsulinemia. J. Clin. Endocrinol. Metab. 2001; 86: 1930–1935.
[29] Yamauchi,T., Kamon,J., Waki,H,, Terauchi,Y., Kubota,N., Hara,K. et al. The fat-derived hormone adiponectin reverses insulin resistance associated with both lipoatrophy and obesity. Nature Med. 2001; 7: 941–946.
[30] Thyfault,J.P., Hedberg,E.M., Anchan,R.M., Thorne,O.P., Isler,C.M., Newton,E.R. et al. Gestational diabetes is associated with depressed adiponectin levels. J. Soc. Gynecol. Invest. 2005; 12: 41–45.
[31] Ramsay,J.E., Jamieson,N., Greer,I.A., Sattar,N. Paradoxical elevation in adiponectin concentrations in women with preeclampsia. Hypertension 2003; 42: 891-894.
[32] O'Sullivan,A.J., Kriketos,A.D., Martin,A. & Brown,M.A. Serum adiponectin levels in normal and hypertensive pregnancy. Hypertens. Preg. 2006; 25: 193-203.
[33] D'Anna,R., Baviera,G., Corrado,F., Giordano,D., Di Benedetto,A. & Jasonni,V.M. Plasma adiponectin concentration in early pregnancy and subsequent risk of hypertensive disorders. Obstet. Gynecol. 2005; 106: 340-344.
[34] Nanda,S., Yu,C.K., Giurcaneanu,L., Akolekar,R. & Nicolaides,K.H. Maternal serum adiponectin at 11-13 weeks of gestation in preeclampsia. Fetal Diagn. Ther. 2011; 29: 208-215.
[35] Lain,K.Y. & Roberts,J.M. Contemporary concepts of the pathogenesis and management of preeclampsia. JAMA 2002; 287: 3183-3186.
[36] Visser,M., Bouter,L.,M., McQuillan,G.M., Wener,M.H. & Harris,T.B. Elevated C reactive protein levels in overweight and obese adults. JAMA 1999; 282: 2131-2135.
[37] Bersinger,N.A., Brandenberger,A.W. & Birkhäuser,M.H. Endometrial and placental protein markers and ovarian steroids in serum during in-vitro fertilisation cycles. Hum. Reprod. 1995; 10, 2149-2154.

[38] Malek,A., Sager,R., Lang,A.B. & Schneider,H. Protein transport across the in vitro perfused human placenta. Am. J. Reprod. Immunol. 1997; 38: 263-271.

[39] Wunder,D.M., Mueller,M.D., Birkhäuser,M.H. & Bersinger,N.A. Increased ENA-78 in the follicular fluid of patients with endometriosis. Acta Obstet. Gynecol. Scand. 2006; 85: 336-342.

[40] Baumann,M.U., Bersinger,N.A., Mohaupt,M.G., Raio,L., Gerber,S. & Surbek,D.V. First-trimester serum levels of soluble endoglin and soluble fms-like tyrosine kinase-1 as first-trimester markers for late-onset preeclampsia. Am. J. Obstet. Gynecol. 2008; 199: 266.e1-266.e6.

[41] Malek,A., Willi,A., Müller,J., Sager,R. & Bersinger,N.A. Capacity for hormone production of cultured trophoblast cells obtained from placentae at term and in early pregnancy. J. Assist. Reprod. Genet. 2001; 18: 299-303.

[42] Maruna,P., Gurlich,R., Frasko,R. & Haluzik,M. Serum leptin levels in septic men correlate well with C-reactive protein (CRP) and TNF-alpha but not with BMI. Physiol. Res. 2001; 50: 589-594.

[43] Hara,K., Horikoshi,M., Yamauchi,T., Yago,H., Miyazaki,O., Ebinuma,H. et al. Measurement of the high-molecular weight form of adiponectin in plasma is useful for the prediction of insulin resistance and metabolic syndrome. Diabetes Care 2006; 29: 1357-1362.

[44] Leung,K.C., Xu,A., Craig,M.E., Martin,A., Lam,K.S. & O'Sullivan,A.J. Adiponectin isoform distribution in women – relationship to female sex steroids and insulin sensitivity. Metabolism 2009; 58: 239-245.

In: Adiponectin: Production, Regulation and Roles in Disease ISBN: 978-1-61942-165-3
Editor: Masayoshi Yamaguchi

Chapter X

CIRCULATING ADIPONECTIN LEVELS IN MALE ADULT PATIENTS: INFLUENCE OF ABDOMINAL OBESITY AND HYPERTRIGLYCERIDEMIA

***T. Meroño*[1], *L. Gómez Rosso*[1], *M. Menafra*[1], *G. Giunta*[2], *L. Boero*[1], *L. Schreier*[1], *L. Cuniberti*[2] and *F. Brites*[1]**[1]

[1]Laboratory of Lipids and Lipoproteins
Department of Clinical Biochemistry
School of Pharmacy and Biochemistry, INFIBIOC
University of Buenos Aires, Argentina
[2]Laboratory of Lipids and Atherosclerosis
Favaloro University, Argentina

ABSTRACT

Recent studies have revealed that adiponectin facilitates insulin action and presents antiatherogenic properties. However, data on the regulation of adiponectin plasma concentration and its relationship with abdominal obesity, with hypertriglyceridemia (HTG) and other metabolic characteristics are still insufficient.

To describe the main factors conditioning plasma adiponectin concentration and to evaluate its association with abdominal obesity and biomarkers of insulin resistance and vascular inflammation in a cohort of male adults attending a health centre.

Male adults attending Favaloro Hospital, Buenos Aires, Argentina, at time of a health control were enrolled. Patients were excluded if they presented diabetes, thyroid, renal or hepatic diseases, previous history of cardiovascular disease, or actual treatment with drugs known to affect carbohydrate or lipid metabolism. Plasma levels of glucose, insulin, triglycerides (TG), total cholesterol, HDL-C, remnant lipoproteins (RLP)-C and LDL-C were measured by standardized methods. CETP activity and the concentrations of adiponectin and endothelial-derived cell adhesion molecules (VCAM-1, ICAM-1 and E-

[1] E-mail address: fdbrites@hotmail.com

selectin) were also determined. Subjects were divided according to WC quartiles: Q1) WC < 90cm (n = 22); Q2) WC 90-98cm (n = 26); Q3) WC 99-103cm (n = 18); and Q4) WC > 103cm (n = 20). Linear trends across WC quartiles were analyzed. Effects of abdominal obesity and HTG (TG>150mg/dl) on adiponectin were evaluated in a linear model adjusted by confounders.

Across WC quartiles, positive linear trends were evidenced for age, hypertensive and HTG subjects, insulin concentration, HOMA-IR, TG, RLP-C, CETP activity and VCAM-1 levels; while, negative ones for HDL-C, RLP-C/TG ratio and adiponectin (all $p < 0.05$). Furthermore, adiponectin levels were correlated with WC ($r = -0.28$, $p < 0.01$), glucose ($r= -0.25$, $p < 0.05$), HOMA-IR ($r = -0.40$, $p < 0.001$), TG ($r = -0.31$, $p < 0.005$), HDL-C ($r = 0.26$, $p < 0.05$), RLP-C ($r = -0.29$, $p < 0.01$), VCAM-1 levels ($r = -0.48$, $p < 0.001$) and CETP activity ($r = -0.28$, $p < 0.01$). Furthermore, abdominal obesity did not exert a significant effect on circulating adiponectin concentration while HTG did ($F = 4.69$, $p < 0.05$). In addition, subjects with HTG displayed lower adiponectin than non-HTG ones, only when they fitted the Q1 and Q2 (4.7 [2.8-6.2] vs. 6.7 [5.3-10.7] μg/ml, $p < 0.001$ for Q1 and Q2 individuals and 4.6 [3. 2-6.0] vs. 5. 1 [3.9-7.1]μg/ml, $p > 0.05$, for Q3 and Q4 subjects). Moreover, adiponectin antiatherogenic properties were evidenced as its concentration explained a 22% of VCAM-1 levels total variance.

Patients with higher grade of abdominal obesity, in addition to increment in markers of insulin resistance, presented higher levels of biomarkers of vascular inflammation and atherosclerosis. Adiponectin plasma levels remained high only in patients with low WC and without HTG. TG concentration might be a better tool than WC for easily identifying subjects with low adiponectin concentration, whose vascular beds might be less protected.

Keywords: adiponectin

INTRODUCTION

Adiponectin is an adipocytokine highly and specifically expressed by adipocytes. Like tumor necrosis factor-α, leptin and resistin, adiponectin is considered an adipose tissue regulatory peptide. This specific-adipose tissue protein was notable by the fact that its plasma concentration, unlike leptin, was found reduced in obese patients in comparison to non-obese subjects [1]. Later on, various studies on adiponectin biological effects added enough evidence to attribute adiponectin antiatherogenic and antiinflammatory properties [2, 3]. Regarding its mechanisms, adiponectin mediates fatty acid metabolism, stimulates glucose uptake [4], modulates glucose metabolism [5] and inhibits endothelium expression of cell adhesion molecules (VCAM-1, ICAM-1 and E-selectin) [3] and foam cell formation [6].

Among the factors known to affect adiponectin expression in human adipocytes, vitamins E [7] and D [8], n-3 polyunsaturated fatty acids [9] and cardiac natriuretic peptides [10] are known to up-regulate adiponectin, while inflammatory mediators [11] and increased adiposity [12] down-regulate it. It is noteworthy, that in most of the aforementioned cases the activation or inactivation of the transcription factor known as peroxisome proliferator-activated receptor γ (PPAR γ) is accountable for the regulation of adiponectin expression [2, 7-9, 11].

In clinical studies, low adiponectin concentration was shown to be a risk factor for type 2 diabetes and cardiovascular disease (CVD) [13-16]. In particular, a recent meta-analysis associated the increase in one logarithmic unit of adiponectin concentration with a 30% reduction in the relative risk for developing type 2 diabetes [13]. Moreover, regarding CVD, low adiponectin was not only identified as an independent risk factor for coronary heart disease in men from the Framingham offspring Study [14], but it was also shown to be associated with chronic heart failure [16] and left ventricular diastolic dysfunction [15]. Overall these results highlight the importance of identifying individuals with low adiponectin concentration who might be more sensitive to develop type 2 diabetes and/or CVD.

Low adiponectin levels have been described in some insulin-resistant states such as obesity (specially abdominal or central obesity) [1], type 2 diabetes [17], lipodystrophy [18], primary hypertriglyceridemia [19] and metabolic syndrome [20], every and each condition associated with high risk of CVD. However, in order to reduce the general burden of CVD, easy strategies to early identify subjects at risk for CVD should be implemented. In this regard, though metabolic syndrome and the hypertriglyceridemic waist phenotype are suggested as conditions which might be sensitive enough to early identify subjects with low adiponectin concentration, their use in CVD prevention still remains under debate [20-22]. In addition, there are scarce studies that evaluate the individual impact of the different components, clustered in the metabolic syndrome definition, over circulating adiponectin levels.

To address this particular issue, the aim of the present chapter was to describe the main factors conditioning plasma adiponectin concentration and to evaluate its association with abdominal obesity and biomarkers of insulin resistance and vascular inflammation in a cohort of male adults attending a health care centre.

METHODS

Subjects

Male adults attending Favaloro Hospital, Buenos Aires, Argentina, at time of a health control were enrolled. Exclusion criteria included: diabetes, thyroid, renal or hepatic diseases, previous history of CVD, or actual treatment with drugs known to affect carbohydrate or lipid metabolism. From all the surveyed individuals, 86 were eligible for the study. None of them showed additional causes of dyslipidemia such as excessive tobacco (> 10 cigarettes/day) or ethanol consumption (> 30 g/day). Total population was divided according to waist circumference (WC) quartiles as Q1) WC < 90cm (n = 22); Q2) WC 90-98cm (n = 26); Q3) WC 99-103cm (n = 18); and Q4) WC > 103cm (n = 20). For sub-analyses, subjects were further classified according to the presence of hypertriglyceridemia (triglycerides ≥ 150 mg/dl) following the National Cholesterol Education Program-Adult Treatment Panel III (ATP III) criteria [23]. Informed consent was obtained from all participants and the protocol was approved by the Ethics Committee from the School of Pharmacy and Biochemistry, University of Buenos Aires and the Institute of Cardiology and Cardiovascular Surgery of the Favaloro Foundation, Buenos Aires, Argentina according to Declaration of Helsinki.

Clinical and Anthropometric Measurements

Age, presence of familial history of cardiovascular disease or diabetes, alcohol consumption and smoking status were recorded during a face-to-face interview. Weight and height were recorded with the subject without shoes and wearing light clothes. Body mass index (BMI) was calculated. Waist circumference was measured at the level of the umbilicus without clothing. Blood pressure was determined after the subject was seated at rest for five minutes employing a mercury sphygmomanometer previously calibrated. Subjects were classified according to the presence of hypertension per ATP III criteria [23]. Every determination was carried out by the same physician.

Biochemical Determination

Blood samples were obtained after a 12-hour overnight fast for biochemical determinations. Glucose, total cholesterol and triglyceride levels were assessed by standardized techniques (Roche Diagnostics, Mannheim, Germany) in a Hitachi 917 (Hitachi, Tokyo, Japan) analyzer. Low-density lipoprotein cholesterol (LDL-C) and high-density lipoprotein cholesterol (HDL-C) concentrations were determined by selective precipitation methods. Remnant-lipoproteins-cholesterol (RLP-C) was calculated as the difference between the cholesterol contained in the supernatants obtained for LDL and HDL measurements [24]. The index RLP-C/TG was calculated to estimate the composition of remnant lipoproteins. Insulin concentration was measured by microparticle enzyme immunoassay (MEIA) (ABBOTT, Japan). The following equation for HOMA-IR index was used: fasting (insulin (μU/ml) x fasting glucose (mmol/l))/22.5. Adiponectin, VCAM-1, ICAM-1 and E-selectin levels were measured by monoclonal antibody based enzyme-linked inmunosorbent assays (ELISA) (R&D Systems, USA)

CETP Activity

CETP activity was determined in serum samples following the general procedure previously described by Lagrost *et al.* [25]. Briefly, the ability of serum to promote the transfer of tritiated cholesteryl esters from a tracer amount of biosynthetically labelled HDL_3 (^{3}H-CE-HDL_3) (NEN Life Science Products, Boston, USA) towards serum apo B-containing lipoproteins was evaluated. Results were expressed as percentage of ^{3}H-cholesteryl esters transferred from HDL_3 to apo B-containing lipoproteins, per ml, per hour. Measurements were all carried out within the same assay. Within-run coefficient of variation (CV) was 5.1%.

Statistical Analyses

Data distribution was tested by Shapiro-Wilks method. Parameters following a Gaussian distribution were presented as the mean ± standard deviation and data following a skewed distribution were presented as median (Q1-Q3). Trends across WC quartiles were assessed using linear contrast by a one-way analysis of variance (ANOVA) [26]. Correlations with adiponectin and VCAM-1 levels were assessed using Spearman correlation test. For linear analyses skewed variables were $\log_e$-transformed. Effects of abdominal obesity (WC quartiles) and HTG on $\log_e$-transformed adiponectin levels were evaluated using a linear model including age, smoking status, hypertension and low HDL-C (HDL-C<40mg/dl) as fixed factors or covariates, correspondingly. Further on, to analyze the associations between vascular inflammation and adiponectin concentration, a stepwise multiple linear regression analysis was carried out employing VCAM-1 as dependent variable and age, smoking status, hypertension, HOMA-IR, triglycerides and HDL-C as independent variables. Analyses were carried out using the SPSS (Chicago, IL) statistical software package SPSS version 17.0® and the software INFOSTAT® (Grupo INFOSTAT, National University of Córdoba, Córdoba, Argentine).

RESULTS

Clinical and biochemical data from the 86 subjects studied are presented in Table 1. There were no differences in age and in the number of smokers or hypertensive subjects ($p > 0.05$). Nonetheless, linear trends across WC quartiles were positive for age and percentages of hypertensive and hypertriglyceridemic subjects. Regarding metabolic biomarkers, as WC increased, insulin concentration, HOMA-IR, triglycerides levels, CETP activity and VCAM-1 concentrations were higher; while HDL-C, RLP-C/TG ratio and adiponectin levels were lower (Table 1 and Figure 1).

Taking into account that as the degree of abdominal obesity increased, the number of subjects with hypertriglyceridemia was higher, the effects of these factors on adiponectin concentration were considered of particular interest. In a linear model which included age, hypertension, hypertriglyceridemia, abdominal obesity and smoking, only hypertriglyceridemia exerted a significant impact on adiponectin concentration at every WC quartile ($F = 4.69$, $p < 0.05$) (Figure 2). In addition, whereas no significant difference was observed in adiponectin levels between individuals from the first and forth WC quartiles, subjects with hypertriglyceridemia displayed lower adiponectin concentration than non hypertriglyceridemic ones only when they fitted the first and second quartiles (4.7[2.8-6.2] *vs.* 6.7[5.3-10.7]μg/ml, $p < 0.001$ for Q1 and Q2 individuals and 4.6 [3.2-6.0] *vs.* 5.1[3.9-7.1]μg/ml, $p > 0.05$, for Q3 and Q4 subjects).

Adiponectin concentration was significantly associated with WC ($r = -0.28$, $p < 0.01$), glucose ($r = -0.25$, $p < 0.05$) and insulin concentrations ($r = -0.40$, $p < 0.001$), HOMA-IR ($r = -0.40$, $p < 0.001$), triglyceride ($r = -0.31$, $p < 0.005$), HDL-C ($r = 0.26$, $p < 0.05$), RLP-C ($r = -0.29$, $p < 0.01$), and VCAM-1 ($r = -0.48$, $p < 0.001$) levels, and CETP activity ($r = -0.28$, $p < 0.01$). As VCAM-1 concentration was the only cell adhesion molecule assayed that showed a linear trend across WC quartiles, its relationship with metabolic parameters was assessed. Higher VCAM-1 levels correlated positively with hypertension ($r = 0.22$, $p < 0.05$), WC ($r = 0.23$, $p < 0.05$) and glucose concentration ($r = 0.25$, $p < 0.05$), while negatively with HDL-C ($r = -0.23$, $p < 0.05$) and adiponectin ($r = -0.48$, $p < 0.001$). Not only adiponectin and VCAM-1 concentrations were significantly correlated, but also by stepwise multiple regression analysis a 22% of VCAM-1 total variance was attributed to adiponectin in a model adjusted by age, hypertension, smoking, HOMA-IR, triglycerides and HDL-C. Further on, when age and smoking were included by the analysis, all together, these variables were able to explain up to 32% of the total variance of VCAM-1 levels (Table 2)

Table 1. Clinical and biochemical characteristics according to waist circumference quartiles

WC (cm)	< 90 (n = 22)	90 ≤ CC < 98cm (n = 26)	98 ≤ CC < 103cm (n = 18)	≥ 103cm (n = 20)	*p* for trend
Age (years)	41 ± 16	45 ± 12	47 ± 10	48 ± 10	0.05
Smokers (%)	14	23	28	10	NS
HTA (%)	23	33	46	60	0.01
Glucose (mg/dl)	91.9 ± 10.4	102.2 ± 22.0	96.2 ± 8.4	99.3 ± 11.3	NS
Insulin (mU/l)	6.6 ± 3.2	9.3 ± 4.3	12.0 ± 6.8	12.0 ± 5.2	0.0001
HOMA-IR	1.51 ± 0.77	2.39 ± 1.27	2.87 ± 1.62	3.00 ± 1.38	0.0001
TG (mg/dl)	118 (90-193)	177 (106-277)	226 (110-312)	202 (144-373)	0.01
HTG (%)	32	61	67	75	0.001
TC (mg/dl)	234 ± 53	247 ± 60	229 ± 48	229 ± 56	NS
RLP-C (mg/dl)	22 (15-39)	31 (19-54)	32 (19-42)	32 (23-48)	0.01
LDL-C (mg/dl)	160 ± 54	156 ± 46	153 ± 38	155 ± 60	NS
HDL-C (mg/dl)	44 ± 13	45 ± 12	40 ± 10	34 ± 9	0.01
RLP-C/TG	0.16 (0.14-0.22)	0.18 (0.15-0.20)	0.15 (0.14-0.18)	0.15 (0.12-0.17)	0.05
CETP (%/ml.h)	257 (232-275)	267 (256-288)	263 (236-279)	280 (261-295)	0.01
VCAM-1 (ng/l)	13 ± 4	16 ± 5	14 ± 5	23 ± 7	0.05
ICAM-1 (ng/l)	15 ± 3	17 ± 5	17 ±3	17 ± 6	NS
E-selectin (ng/l)	24 (15-37)	25 (15-37)	22 (19-44)	21 (14-32)	NS

HTA, hypertension; HTG, hypertriglyceridemic; HOMA-IR, Homeostasis Model Assessment; TG, triglycerides; TC, total cholesterol; RLP-C, remnant lipoproteins-cholesterol; LDL, low density lipoproteins; HDL, high density lipoproteins; VCAM, vascular cell adhesion molecule. Results were expressed as media±S.D. or median (range), depending on parametric or non parametric distribution of the data, respectively.

Dependent variable, VCAM-1 concentration. WC, waist circumference; HOMA-IR, Homeostasis model assessment; HDL, high density lipoprotein.

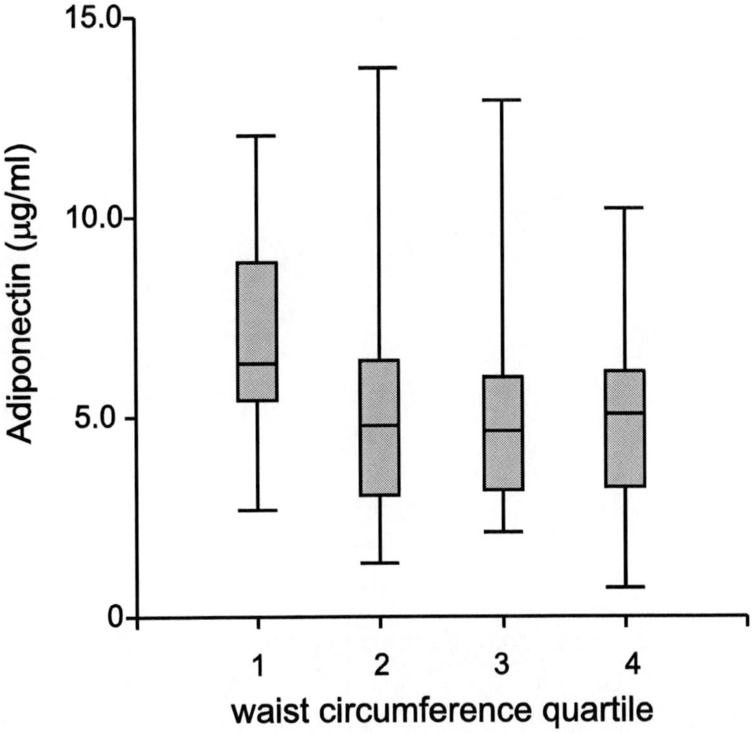

Figure 1. Adiponectin concentration across waist circumference quartiles. Quartiles were defined as: Q1) WC < 90cm (n = 22); Q2) WC 90-98cm (n = 26); Q3) WC 99-103cm (n = 18); and Q4) WC > 103cm (n = 20). $p < 0.01$ for linear trend.

Table 2. Multiple linear regression analysis for the association of VCAM-1 concentration as dependent variable

Model	B	Std. Error	t	Significance	R^2
Adiponectin	-7.86	1.38	-5.7	0.001	0.32
Age	0.25	0.08	3.1	0.005	
Smoking	-7.64	2.70	-2.8	0.01	
Hypertension	1.91	2.15	0.9	NS	
WC	0.09	0.12	0.7	NS	
HOMA-IR	-0.41	0.79	-0.5	NS	
Triglycerides	-2.55	1.84	-1.4	NS	
HDL-C	-0.13	0.10	-1.3	NS	

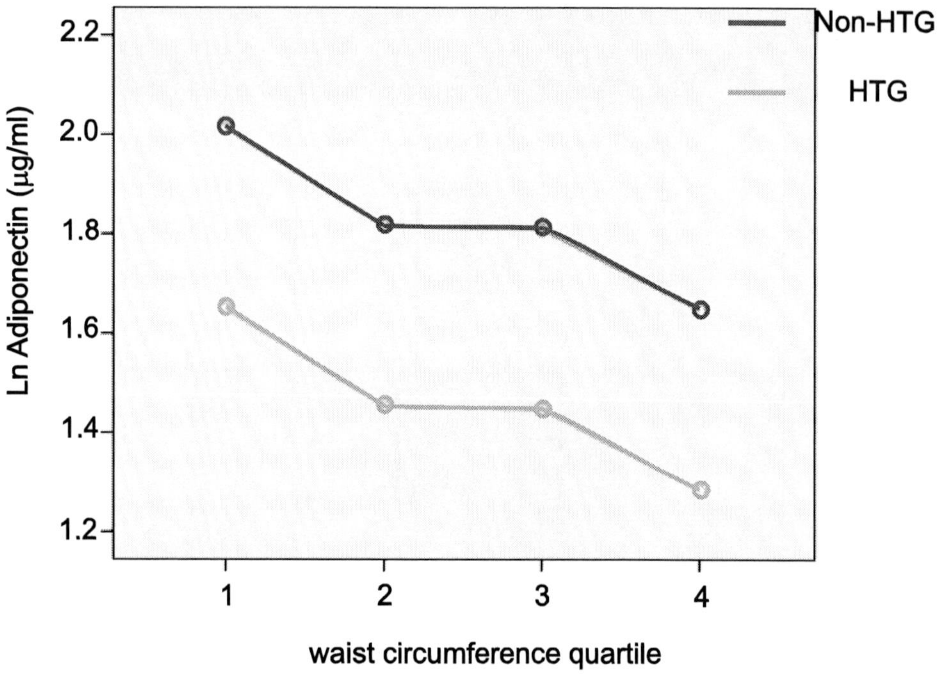

Figure 2. Marginal means of $\log_e$-transformed adiponectin concentration according to waist circumference quartiles and to the presence of hypertriglyceridemia. Means were calculated in a model which included age, smoking, hypertension and low HDL-C. Quartiles were defined as: Q1) WC < 90cm (n = 22); Q2) WC 90-98cm (n = 26); Q3) WC 99-103cm (n = 18); and Q4) WC > 103cm (n = 20).

DISCUSSION

The present chapter identified hypertriglyceridemia as one of the most important factors contributing to the decrease in circulating adiponectin concentration in male subjects. Though adiponectin decreased across WC quartiles, suggesting an effect of abdominal obesity, it did not reach statistical significance in the fully-adjusted model. These results support the clinical importance of early identifying subjects with high triglyceride concentration (triglycerides > 150 mg/dl) as hypertriglyceridemia was more sensitive to detect individuals with decreased adiponectin concentration. Moreover, the effect of hypertriglyceridemia was more pronounced on subjects with low WC, even in those with WC below 90 cm, which is the cut-off point established for males in the definition of both metabolic syndrome [27] and the hypertriglyceridemic waist phenotype [28]. In fact, a recent study revealed that these definitions not only present an important overlap in subject classification, but also the plasma concentrations of various cardiometabolic risk biomarkers, including adiponectin, were similar disregarding the classification criteria employed [20].

In agreement with previous studies, patients with higher grade of abdominal obesity, besides lower adiponectin concentration, displayed an increase in insulin resistance biomarkers, cardiometabolic risk factors and higher levels of biomarkers of vascular inflammation. In this aspect, the antiatherogenic role of adiponectin was highlighted by its negative association with VCAM-1 levels in the univariate and multivariate analyses. Adiponectin was able to explain up to 22% of VCAM-1 total variance, and in conjunction with age and smoking, it reached 32%, in a model adjusted by hypertension, HOMA-IR, triglycerides and HDL-C. This intimate association might be reflect of the capacity of adiponectin to prevent endothelial expression of cell adhesion molecules, and therefore, atherosclerosis development [3]. In the present chapter, in agreement with previous reports [29,30], it was remarkable that the higher WC, the higher VCAM-1 concentration was.

Cell adhesion molecule levels had been associated with several clinical presentations of cardiovascular disease [31]. Moreover, its levels were also shown to be increased in several conditions known to be associated to high risk of cardiovascular disease, like metabolic syndrome and type 2 diabetes [32-34]. In this regard, cell adhesion molecules were also shown to be a risk factor for CVD [35-37]. Particularly, a prospective study among patients with coronary artery disease associated high VCAM-1 levels with a 2.8- fold increase in the risk for future fatal cardiovascular events in a fully-adjusted model by other inflammatory markers [38]. Moreover, VCAM-1 improved the risk prediction of death from CVD in patients presenting acute coronary syndromes [37]. Regarding the mechanism underlying this association, it has been established that under certain circumstances that trigger vascular inflammation [39, 40], endothelial cells express cell adhesion molecules that home monocytes and inflammatory cells into the intima. Then, these cells secrete various inflammatory cytokines and increase the local concentration of oxidative products, perpetuating vascular damage and initiating the early atherosclerotic lesion [41, 42].

Obesity, in particular abdominal obesity, is largely known to be associated with insulin resistance and with alterations in lipoprotein concentration and quality [32,43]. In this respect, remnant lipoproteins were more abundant and more enriched in triglycerides as WC increased. Remnant lipoproteins were shown to facilitate atherosclerosis development and various studies associated its plasma concentration with increased risk of CVD [24, 44, 45]. Furthermore, it has been reported that as remnant lipoproteins increased their triglyceride content, their capacity to induce vascular inflammation increases and, consequently, CVD risk [46, 47]. In the present chapter, adiponectin was negatively associated with RLP-C, thus suggesting that adiponectin might stimulate triglyceride rich lipoprotein catabolism. Even if the relationship between adiponectin and insulin resistance has been clearly established, the inverse correlations between adiponectin and triglyceride, HDL-C or small and dense LDL levels were reported to be independent of the amount of intraabdominal fat and the degree of insulin-resistance [48, 49]. Moreover, Chan et al. [50] proposed that adiponectin levels would exert an independent role in regulating the metabolism of triglyceride-rich lipoproteins, evaluated through the measurement of plasma concentrations of apolipoprotein (apo) B48, apo C-III, and the cholesterol transported in remnant-like particles.

As RLP triglyceride content is known to be a determinant factor for CETP activity, this direct association between adiponectin and lipoprotein metabolism could be, in part, mediated by CETP modulation. Actually, adiponectin was inversely correlated with adiponectin concentration. CETP physiological role consists of interchanging triglycerides by cholesteryl esters between lipoproteins, thus generating cholesterol-enriched and triglyceride-depleted apo B-containing lipoproteins. This modification introduced by CETP could slow their catabolism as it has been demonstrated that triglyceride content is determinant for lipoprotein lipase activity [51]. Moreover, CETP inhibition was demonstrated to decrease RLP atherogenicity by decreasing its concentration and improving its quality [52]. Overall, as decreasedadiponectin concentration, accumulation of RLP and increased CETP activity might all together contribute to the development of atherosclerosis; the early identification of the individuals with low adiponectin concentration represents a topic of clinical importance.

In conclusion, patients with higher grade of abdominal obesity, in addition to increment in markers of insulin resistance, presented higher levels of biomarkers of vascular inflammation and atherosclerosis. Adiponectin plasma levels remained high only in patients with low WC and without hypertriglyceridemia. Triglyceride concentration might be a better tool than WC for easily identifying subjects with low adiponectin concentration. Data form univariate and multivariate analyses carried out in the present chapter reaffirm the multiple antiatherogenic properties of adiponectin.

REFERENCES

[1] Y. Arita, S. Kihara, N. Ouchi, M. Takahashi, K. Maeda, J. Miyagawa, K. Hotta, I. Shimomura, T. Nakamura, K. Miyaoka, H. Kuriyama, M. Nishida, S. Yamashita, K. Okubo, K. Matsubara, M. Muraguchi, Y. Ohmoto, T. Funahashiand Y. Matsuzawa. Paradoxical decrease of an adipose-specific protein, adiponectin, in obesity. *Biochem Biophys Res Commun*. 257, 79-83 (1999).

[2] K. R. Rabin, Y. Kamari, I. Avni, E. Grossmanand Y. Sharabi. Adiponectin: linking the metabolic syndrome to its cardiovascular consequences. *Expert Rev Cardiovasc Ther*. 3, 465-71 (2005).

[3] P. Zhang, Y. Wang, Y. Fan, Z. Tangand N. Wang. Overexpression of adiponectin receptors potentiates the antiinflammatory action of subeffective dose of globular adiponectin in vascular endothelial cells. *Arterioscler Thromb Vasc Biol*. 29, 67-74 (2009).

[4] T. Yamauchi, J. Kamon, Y. Minokoshi, Y. Ito, H. Waki, S. Uchida, S. Yamashita, M. Noda, S. Kita, K. Ueki, K. Eto, Y. Akanuma, P. Froguel, F. Foufelle, P. Ferre, D. Carling, S. Kimura, R. Nagai, B. B. Kahnand T. Kadowaki. Adiponectin stimulates glucose utilization and fatty-acid oxidation by activating AMP-activated protein kinase. *Nat Med*. 8, 1288-95 (2002).

[5] A. H. Berg, T. P. Combs, X. Du, M. Brownleeand P. E. Scherer. The adipocyte-secreted protein Acrp30 enhances hepatic insulin action. *Nat Med*. 7, 947-53 (2001).

[6] K. Tsubakio-Yamamoto, F. Matsuura, M. Koseki, H. Oku, J. C. Sandoval, M. Inagaki, K. Nakatani, H. Nakaoka, R. Kawase, M. Yuasa-Kawase, D. Masuda, T. Ohama, N. Maeda, Y. Nakagawa-Toyama, M. Ishigami, M. Nishida, S. Kihara, I. Shimomuraand S. Yamashita. Adiponectin prevents atherosclerosis by increasing cholesterol efflux from macrophages. *Biochem Biophys Res Commun*. 375, 390-4 (2008).

[7] J. F. Landrier, E. Gouranton, C. El Yazidi, C. Malezet, P. Balaguer, P. Boreland M. J. Amiot. Adiponectin expression is induced by vitamin E via a peroxisome proliferator-activated receptor gamma-dependent mechanism. *Endocrinology*. 150, 5318-25 (2009).

[8] S. Lorente-Cebrian, A. Eriksson, T. Dunlop, N. Mejhert, I. Dahlman, G. Astrom, E. Sjolin, K. Wahlen, C. Carlberg, J. Laurencikiene, P. Heden, P. Arnerand M. Ryden.

Differential effects of 1alpha,25-dihydroxycholecalciferol on MCP-1 and adiponectin production in human white adipocytes. *Eur J Nutr*. (2011).

[9] M. Itoh, T. Suganami, N. Satoh, K. Tanimoto-Koyama, X. Yuan, M. Tanaka, H. Kawano, T. Yano, S. Aoe, M. Takeya, A. Shimatsu, H. Kuzuya, Y. Kameiand Y. Ogawa. Increased adiponectin secretion by highly purified eicosapentaenoic acid in rodent models of obesity and human obese subjects. *Arterioscler Thromb Vasc Biol*. 27, 1918-25 (2007).

[10] C. Moro, E. Klimcakova, K. Lolmede, M. Berlan, M. Lafontan, V. Stich, A. Bouloumie, J. Galitzky, P. Arnerand D. Langin. Atrial natriuretic peptide inhibits the production of adipokines and cytokines linked to inflammation and insulin resistance in human subcutaneous adipose tissue. *Diabetologia*. 50, 1038-47 (2007).

[11] G. Zappalaand M. M. Rechler. IGFBP-3, hypoxia and TNF-alpha inhibit adiponectin transcription. *Biochem Biophys Res Commun*. 382, 785-9 (2009).

[12] A. T. Turer, A. Khera, C. R. Ayers, C. B. Turer, S. M. Grundy, G. L. Vegaand P. E. Scherer. Adipose tissue mass and location affect circulating adiponectin levels. *Diabetologia*. (2011).

[13] S. Li, H. J. Shin, E. L. Dingand R. M. van Dam. Adiponectin levels and risk of type 2 diabetes: a systematic review and meta-analysis. *Jama*. 302, 179-88 (2009).

[14] M. Ai, S. Otokozawa, B. F. Asztalos, C. C. White, L. A. Cupples, K. Nakajima, S. Lamon-Fava, P. W. Wilson, Y. Matsuzawaand E. J. Schaefer. Adiponectin: An independent risk factor for coronary heart disease in men in the Framingham offspring Study. *Atherosclerosis*. 217, 543-8 (2011).

[15] H. Fukuta, N. Ohte, K. Wakami, T. Goto, T. Taniand G. Kimura. Relation of Plasma Levels of Adiponectin to Left Ventricular Diastolic Dysfunction in Patients Undergoing Cardiac Catheterization for Coronary Artery Disease. *Am J Cardiol*. (2011).

[16] T. Tsutamoto, T. Tanaka, H. Sakai, C. Ishikawa, M. Fujii, T. Yamamotoand M. Horie. Total and high molecular weight adiponectin, haemodynamics, and mortality in patients with chronic heart failure. *Eur Heart J*. 28, 1723-30 (2007).

[17] K. Hotta, T. Funahashi, Y. Arita, M. Takahashi, M. Matsuda, Y. Okamoto, H. Iwahashi, H. Kuriyama, N. Ouchi, K. Maeda, M. Nishida, S. Kihara, N. Sakai, T. Nakajima, K. Hasegawa, M. Muraguchi, Y. Ohmoto, T. Nakamura, S. Yamashita, T. Hanafusaand Y. Matsuzawa. Plasma concentrations of a novel, adipose-specific protein, adiponectin, in type 2 diabetic patients. *Arterioscler Thromb Vasc Biol*. 20, 1595-9 (2000).

[18] W. A. Haque, I. Shimomura, Y. Matsuzawaand A. Garg. Serum adiponectin and leptin levels in patients with lipodystrophies. *J Clin Endocrinol Metab*. 87, 2395 (2002).

[19] L. Gomez Rosso, T. Merono, M. B. Benitez, G. Lopez, G. Giunta, M. L. D'Ambrosio, R. Wikinski, L. Cunibertiand F. Brites. Low adiponectin levels in primary hypertriglyceridemic male patients. *Nutr Metab Cardiovasc Dis*. 19, 135-9 (2009).

[20] P. Blackburn, I. Lemieux, N. Almeras, J. Bergeron, M. Cote, A. Tremblay, B. Lamarcheand J. P. Despres. The hypertriglyceridemic waist phenotype versus the National Cholesterol Education Program-Adult Treatment Panel III and International Diabetes Federation clinical criteria to identify high-risk men with an altered cardiometabolic risk profile. *Metabolism*. 58, 1123-30 (2009).

[21] G. M. Reaven. The metabolic syndrome: time to get off the merry-go-round? *J Intern Med*. 269, 127-36 (2011).

[22] M. R. Salazar, H. A. Carbajal, W. G. Espeche, C. A. Dulbecco, M. Aizpurua, A. G. Marillet, R. F. Echeverriaand G. M. Reaven. Relationships among insulin resistance, obesity, diagnosis of the metabolic syndrome and cardio-metabolic risk. *Diab Vasc Dis Res*. 8, 109-16 (2011).

[23] ATP III. Third Report of the National Cholesterol Education Program (NCEP) Expert Panel on Detection, Evaluation, and Treatment of High Blood Cholesterol in Adults (Adult Treatment Panel III) final report. *Circulation*. 106, 3143-421 (2002).

[24] B. G. Nordestgaard, M. Benn, P. Schnohrand A. Tybjaerg-Hansen. Nonfasting triglycerides and risk of myocardial infarction, ischemic heart disease, and death in men and women. *Jama*. 298, 299-308 (2007).

[25] L. Lagrost, H. Gandjini, A. Athias, V. Guyard-Dangremont, C. Lallemantand P. Gambert. Influence of plasma cholesteryl ester transfer activity on the LDL and HDL distribution profiles in normolipidemic subjects. *Arterioscler Thromb*. 13, 815-25 (1993).

[26] V. Bewick, L. Cheekand J. Ball. Statistics review 9: one-way analysis of variance. *Crit Care*. 8, 130-6 (2004).

[27] K. G. Alberti, R. H. Eckel, S. M. Grundy, P. Z. Zimmet, J. I. Cleeman, K. A. Donato, J. C. Fruchart, W. P. James, C. M. Loriaand S. C. Smith, Jr. Harmonizing the metabolic syndrome: a joint interim statement of the International Diabetes Federation Task Force on Epidemiology and Prevention; National Heart, Lung, and Blood Institute; American Heart Association; World Heart Federation; International Atherosclerosis Society; and International Association for the Study of Obesity. *Circulation*. 120, 1640-5 (2009).

[28] I. Lemieux, A. Pascot, C. Couillard, B. Lamarche, A. Tchernof, N. Almeras, J. Bergeron, D. Gaudet, G. Tremblay, D. Prud'homme, A. Nadeauand J. P. Despres. Hypertriglyceridemic waist: A marker of the atherogenic metabolic triad (hyperinsulinemia; hyperapolipoprotein B; small, dense LDL) in men? *Circulation*. 102, 179-84 (2000).

[29] K. M. Pou, J. M. Massaro, U. Hoffmann, R. S. Vasan, P. Maurovich-Horvat, M. G. Larson, J. F. Keaney, Jr., J. B. Meigs, I. Lipinska, S. Kathiresan, J. M. Murabito, C. J. O'Donnell, E. J. Benjaminand C. S. Fox. Visceral and subcutaneous adipose tissue volumes are cross-sectionally related to markers of inflammation and oxidative stress: the Framingham Heart Study. *Circulation*. 116, 1234-41 (2007).

[30] S. Kapiotis, G. Holzer, G. Schaller, M. Haumer, H. Widhalm, D. Weghuber, B. Jilma, G. Roggla, M. Wolzt, K. Widhalmand O. F. Wagner. A proinflammatory state is detectable in obese children and is accompanied by functional and morphological vascular changes. *Arterioscler Thromb Vasc Biol*. 26, 2541-6 (2006).

[31] H. H. Lu, Z. Q. Sheng, Y. Wangand L. Zhang. Levels of soluble adhesion molecules in patients with various clinical presentations of coronary atherosclerosis. *Chin Med J (Engl)*. 123, 3123-6 (2010).

[32] L. Gomez Rosso, M. B. Benitez, M. C. Fornari, V. Berardi, S. Lynch, L. Schreier, R. Wikinski, L. Cunibertiand F. Brites. Alterations in cell adhesion molecules and other biomarkers of cardiovascular disease in patients with metabolic syndrome. *Atherosclerosis*. 199, 415-23 (2008).

[33] I. Kowalska, M. Straczkowski, M. Szelachowska, I. Kinalska, J. Prokop, H. Bachorzewska-Gajewskaand A. Stepien. Circulating E-selectin, vascular cell adhesion molecule-1, and intercellular adhesion molecule-1 in men with coronary artery disease assessed by angiography and disturbances of carbohydrate metabolism. *Metabolism*. 51, 733-6 (2002).

[34] P. Marques-Vidal, E. Mazoyer, V. Bongard, P. Gourdy, J. B. Ruidavets, L. Drouetand J. Ferrieres. Prevalence of insulin resistance syndrome in southwestern France and its relationship with inflammatory and hemostatic markers. *Diabetes Care*. 25, 1371-7 (2002).

[35] E. B. Jude, J. T. Douglas, S. G. Anderson, M. J. Youngand A. J. Boulton. Circulating cellular adhesion molecules ICAM-1, VCAM-1, P- and E-selectin in the prediction of cardiovascular disease in diabetes mellitus. *Eur J Intern Med*. 13, 185-189 (2002).

[36] C. Schmidt, J. Hultheand B. Fagerberg. Baseline ICAM-1 and VCAM-1 are increased in initially healthy middle-aged men who develop cardiovascular disease during 6.6 years of follow-up. *Angiology*. 60, 108-14 (2009).

[37] A. Silvestro, G. Brevetti, V. Schiano, F. Scopacasaand M. Chiariello. Adhesion molecules and cardiovascular risk in peripheral arterial disease. Soluble vascular cell adhesion molecule-1 improves risk stratification. *Thromb Haemost*. 93, 559-63 (2005).

[38] S. Blankenberg, H. J. Rupprecht, C. Bickel, D. Peetz, G. Hafner, L. Tiretand J. Meyer. Circulating cell adhesion molecules and death in patients with coronary artery disease. *Circulation*. 104, 1336-42 (2001).

[39] W. L. Jin, K. Azuma, T. Mita, H. Goto, A. Kanazawa, T. Shimizu, F. Ikeda, Y. Fujitani, T. Hirose, R. Kawamoriand H. Watada. Repetitive hypoglycaemia increases serum adrenaline and induces monocyte adhesion to the endothelium in rat thoracic aorta. *Diabetologia*. 54, 1921-9 (2011).

[40] S. Bansilal, M. E. Farkouhand V. Fuster. Role of insulin resistance and hyperglycemia in the development of atherosclerosis. *Am J Cardiol*. 99, 6B-14B (2007).

[41] R. Stockerand J. F. Keaney, Jr. Role of oxidative modifications in atherosclerosis. Physiol Rev. 84, 1381-478 (2004).

[42] M. G. Ionita, P. van den Borne, L. M. Catanzariti, F. L. Moll, J. P. de Vries, G. Pasterkamp, A. Vinkand D. P. de Kleijn. High neutrophil numbers in human carotid atherosclerotic plaques are associated with characteristics of rupture-prone lesions. Arterioscler Thromb Vasc Biol. 30, 1842-8 (2010).

[43] D. C. Goff, Jr., R. B. D'Agostino, Jr., S. M. Haffnerand J. D. Otvos. Insulin resistance and adiposity influence lipoprotein size and subclass concentrations. Results from the Insulin Resistance Atherosclerosis Study. Metabolism. 54, 264-70 (2005).

[44] K. Nakajima, Y. Nakajima, S. Takeichiand M. Q. Fujita. Plasma remnant-like lipoprotein particles or LDL-C as major pathologic factors in sudden cardiac death cases. Atherosclerosis. 198, 237-46 (2008).

[45] M. Taguchi, M. Ishigami, M. Nishida, T. Moriyama, S. Yamashitaand T. Yamamura. Remnant lipoprotein-cholesterol is a predictive biomarker for large artery atherosclerosis in apparently healthy women: usefulness as a parameter for annual health examinations. Ann Clin Biochem. 48, 332-7 (2011).

[46] J. Dalla-Riva, E. Garonna, J. Elliott, K. M. Bothamand C. P. Wheeler-Jones. Endothelial cells as targets for chylomicron remnants. *Atheroscler Suppl.* 11, 31-7 (2011).

[47] P. N. Hopkins, M. N. Nanjee, L. L. Wu, M. G. McGinty, E. A. Brinton, S. C. Huntand J. L. Anderson. Altered composition of triglyceride-rich lipoproteins and coronary artery disease in a large case-control study. *Atherosclerosis*. 207, 559-66 (2009).

[48] M. Cnop, P. J. Havel, K. M. Utzschneider, D. B. Carr, M. K. Sinha, E. J. Boyko, B. M. Retzlaff, R. H. Knopp, J. D. Brunzelland S. E. Kahn. Relationship of adiponectin to body fat distribution, insulin sensitivity and plasma lipoproteins: evidence for independent roles of age and sex. *Diabetologia*. 46, 459-69 (2003).

[49] R. Baratta, S. Amato, C. Degano, M. G. Farina, G. Patane, R. Vigneriand L. Frittitta. Adiponectin relationship with lipid metabolism is independent of body fat mass: evidence from both cross-sectional and intervention studies. *J Clin Endocrinol Metab*. 89, 2665-71 (2004).

[50] D. C. Chan, G. F. Watts, T. W. Ng, Y. Uchida, N. Sakai, S. Yamashitaand P. H. Barrett. Adiponectin and other adipocytokines as predictors of markers of triglyceride-rich lipoprotein metabolism. *Clin Chem*. 51, 578-85 (2005).

[51] L. Schreier, G. Berg, V. Zago, A. I. Gonzalezand R. Wikinski. Kinetics of in vitro lipolysis of human very low-density lipoprotein by lipoprotein lipase. *Nutr Metab Cardiovasc Dis*. 12, 13-8 (2002).

[52] M. Guerin, W. Le Goff, E. Duchene, Z. Julia, T. Nguyen, T. Thuren, C. L. Shearand M. J. Chapman. Inhibition of CETP by torcetrapib attenuates the atherogenicity of postprandial TG-rich lipoproteins in type IIB hyperlipidemia. *Arterioscler Thromb Vasc Biol*. 28, 148-54 (2008).

INDEX

D

E

F

G

H

I

J

K

L

M

N

O

P

T

U

V

W

X

Y

Z